Oswald Crawfurd

A Year of Sport and Natural History

Shooting, Hunting, Coursing, Falconry and Fishing with Chapters on Birds of Prey, the Nidification of Birds and the Habits of British Wild Birds and Animals

Oswald Crawfurd

A Year of Sport and Natural History
Shooting, Hunting, Coursing, Falconry and Fishing with Chapters on Birds of Prey, the Nidification of Birds and the Habits of British Wild Birds and Animals

ISBN/EAN: 9783337026653

Printed in Europe, USA, Canada, Australia, Japan

Cover: Foto ©berggeist007 / pixelio.de

More available books at **www.hansebooks.com**

AND

NATURAL HISTORY

*SHOOTING, HUNTING, COURSING, FALCONRY
AND FISHING*

WITH CHAPTERS ON

BIRDS OF PREY, THE NIDIFICATION OF BIRDS
AND THE HABITS OF BRITISH WILD
BIRDS AND ANIMALS

EDITED BY

OSWALD CRAWFURD

WITH NUMEROUS ILLUSTRATIONS

BY

FRANK FELLER, BRYAN HOOK, CECIL ALDIN, A. T. ELWES, E. NEALE,
JOHN BEER, P. VIENZENY, STANLEY BERKELEY,
AND G. E. LODGE

LONDON—CHAPMAN AND HALL, Limited
1895

PREFACE.

WHILE superintending the literary department of *Black and White*, I prevailed upon a number of competent writers on Sport and Natural History to deal week by week, the whole of one year round, with these two topics. We called the series "Field Sports and Field Studies," and as the writers knew their business and how to write upon it, the papers were exceedingly popular.

Among the authors were such sportsmen and naturalists as Mr. Aubyn Trevor-Battye, who has since achieved fame in Arctic lands, Mr. George Lindesay, a classic in sporting literature, Mr. H. H. S. Pearse, the admirable writer and referee on hunting of all kinds, and Mr. Sachs, a known and notable angler and writer on angling, with many others of high local or special authority as sportsmen or naturalists. As regards the artists engaged on this book their work is all so praiseworthy that I hardly like to pick and choose from the list, and elect, instead, to print their names in full on our title-page.

I claim for this book, in forty-five parts or sections, that it is not a mere jumbled collection of articles, but a consecutive work dealing, in their due sequence, with forty-five consecutive and most interesting seasonal phases of Sport and Natural History in the British Islands.

OSWALD CRAWFURD, *Editor.*

January, 1895.

CONTENTS.

JANUARY.
	PAGE
Fox-Hunting in the Shires	1
Fox-Hunting Outside the Shires	9
Snipe Shooting	18
Wild Swan Shooting	25
Hunting with Beagles	33

FEBRUARY.
The Wild Goose	41
Rabbit Shooting	49
Spring Salmon Fishing	56

MARCH.
Our Birds of Prey. I. The Owls	64
Our Birds of Prey. II. Hawks, Buzzards, Kites and Harriers	72
Our Birds of Prey. III. Eagles, Falcons and Osprey	82

APRIL.
Bird Nesting. I. Sea-Birds	92
Bird Nesting. II. Moor Birds	104
Bird Nesting. III. Tree Nesting Birds	111
Trout Fishing in Mountain Streams	118

MAY.
Thames Trout Fishing	126
The Tricks of Poachers	133
Fishing with the Dry Fly	142

CONTENTS.

JUNE.

	PAGE
Scotch Loch Fishing	146
Bass-Fishing	153

JULY.

Otter Hunting.	158
Sea Fishing from Piers.	164

AUGUST.

The White Trout.	172
Chub Fishing.	178
Char Fishing.	187
The Habits of the Wild Red Deer	195
Flapper Shooting	200

SEPTEMBER.

Partridge Shooting.	207
Hunting the Wild Red Deer	212
Rabbit Hawking	219

OCTOBER.

Pheasant Shooting.	227
Cub-Hunting	235
Partridge Hawking.	243
Rough Shooting	251

NOVEMBER.

Chantrey's Famous Shot.	259
Tweed Salmon Fishing.	266
Hare Hunting on the Brighton Downs	271
Duck-Shooting on the Broads	283
Coursing.	287
Roe Shooting.	294

DECEMBER.

December Sport in the Highlands	302
A Cock Drive in Scotland.	310
'Longshore Shooting	315
Gamekeepers	320
Pike Fishing	327

LIST OF ILLUSTRATIONS.

	PAGE
The Wild Red Deer	*Frontispiece*
Hunting in the Shires	3
Hunting outside the Shires	11
Snipe Shooting	21
Wild Swan Shooting	27
Hunting with Beagles	35
Flight Shooting	43
Rabbit Shooting	51
Tawny Owl	64
Barn Owl	65
Long-eared Owl	67
Short-eared Owl	70
The Sparrow-Hawk (Accipiter Nisus)	73
The Common Buzzard (Buteo Vulgaris)	75
The Kite (Milvus Ictinus)	77
The Hen-Harrier (Circus Cyaneus)	78
The Marsh-Harrier (Circus Æruginosus)	79
The Golden Eagle	83
The Osprey	85
The Peregrine	87
The Hobby	89
The Kestrel	90
The Tern	93
The Lesser Gull	95
The Puffin	97
Kittiwakes	99
Guillemots on the Needle Rock, Lundy Island	101

LIST OF ILLUSTRATIONS.

	PAGE
The Common Peewit, or Lapwing	105
The Woodcock	107
The Great Northern Diver	109
The Rook	112
The Heron	114
The Woodpecker	116
Trout Fishing in Mountain Streams	121
Thames Trout Fishing	129
Tricks of Poachers	137
Loch Fishing	149
Bass-fishing	155
Otter Hunting	161
Sea Fishing from Piers	167
White Trout Fishing	175
Chub Fishing	181
Char Fishing	191
Flapper Shooting	203
Hunting the Wild Red Deer	215
Rabbit Hawking	221
Pheasants	229
Cub-hunting	239
Partridge Hawking	247
Rough Shooting	255
Chantrey's Famous Shot	261
Tweed Salmon Fishing	269
Hare Hunting on the Brighton Downs	279
Duck-Shooting on the Broads	285
Coursing	289
Roe Shooting	297
December Sport in the Highlands	305
A Would-be Poacher	323

A YEAR OF SPORT
AND
NATURAL HISTORY.

JANUARY.

FOX-HUNTING IN THE SHIRES.

BY H. H. S. PEARSE.

PACE, with all the wild rapture that word conveys, and nothing else, has made midland pastures the favourite hunting-grounds of English men and women who have means to justify and health to enjoy the perfection of pleasure in field sports. None but those who have hunted with the Quorn or Pytchley, the Cottesmore, Billesden, or Belvoir in their best country, when all Melton's bravest and fairest are inspired by keen rivalry, can know what that delight is. The merry men of Lord Eglinton's Hunt will say that their wide stretches of Ayrshire grass lands hold a better scent, that their Foxes are stouter than any to be found in Leicestershire, and that their hounds run as fast; yet they know no crowds like those that congregate at a fixture near Melton Mowbray or Market Harborough. Followers of the Blackmore Vale have been known to hold very similar views as to the claims of

Dorsetshire; and the fox-hunters of many countries that are classed as "provincial" refuse to admit the superiority of the shires in any point worthy to command the admiration of true sportsmen. There is one hunt—in the West—certainly not less distinguished for historic associations or the social celebrity of those who assemble at its fixtures than any in the Midlands. The "blue and bluff" of Badminton must be known to fame all the world over; Dukes of Beaufort for many generations have been acknowledged leaders in the hunting world, and the "badger-pied beauties" as they skim over the open in full chorus, or spread like a rocket to recover the lost scent for themselves or stoop to it with a joyous whimper after Lord Worcester has lifted them forward in one of his masterly casts, make a picture that might well impress the imagination of the coldest critics. Their followers are to be numbered by hundreds, and yet sport with them differs essentially from that of the shires. The Warwickshire hounds are of the best also, and can go fast enough, as every man must confess who ever tried to hold his own with them, when Lord Willoughby cheered "the dappled darlings" over the strongly-fenced pastures of Shuckburgh Vale. This hunt is nearest to the shires geographically and in methods of sport, yet fashion has not admitted it to membership of the guild, probably because Lord Willoughby—though he loves pace no less than the hottest blooded Meltonian—will be master of his own pack, and will not let a hard-riding field dictate terms to him.

It may be granted that all hounds go fast enough at times to run clean away from horsemen, but a brilliant episode of this kind is not quite what one means by pace in the Meltonian sense, which depends rather less on scent than on system. A certain

HUNTING IN THE SHIRES.

happy combination in the first place doubtless gave special distinction to Leicestershire, and the counties bordering on it. No ploughshares at that time fretted the fair surface of those old pastures. For miles in every direction there was nothing to interfere with the dash of a clamorous pack in hot pursuit. Nearly all the fields afforded good galloping ground, and the fences, though few, were big enough to add a strong spice of danger to other pleasures of the chase. These were natural advantages, however, of which the Midlands had not a monopoly, and they alone would not have made Melton the hunting centre of the world. But when the great Hugo Meynell began, at Quorndon Hall, to breed hounds for speed, and to handle them with a quickness that had been previously unknown, the fame of Quorn sport attracted aspiring youths from all quarters of England. They vied with each other in hard riding, until their keen rivalry became an embarrassment to the master and a danger to his hounds. Then the system that had been adopted for pleasure became perpetuated and intensified as a necessity. No huntsman or hound that showed a tendency to dwell upon the line could be tolerated longer. The avalanche of horsemen would be upon them the moment they hung about, and so they got into a habit of flinging forward with a dash whenever scent failed. To have a pack that would bear to be lifted without becoming wild, even amid a bewildering maze of hoofs in rapid motion, became imperative then. The new conditions demanded quick resolution and magnetic power of command in a huntsman ; high courage and the quality known as "drive" in hounds. These were developed in time, and by scientific system they have been transmitted from generation to generation. Nowhere out of the shires does one see a similar display of eager, restless activity all round. The mighty rush of squadrons, where every horseman strives for a lead, apparently reckless of what may befall himself or anything that comes in his way, would seem the very ecstasy of madness to one who had never played a part in such stirring scenes. Steeple-chasing, some critics say it is, but perhaps they are moved to envy by the thought that they can no longer hold their own with the boldest. He must be cold, indeed, whose blood is not stirred to rapture, or who can stop to criticize when Tom Firr, with clear, shrill blasts of the horn that Quornites know so well, is getting his hounds out of Cream Gorse on the line of a Fox whose neck is set straight for the glorious vale of Twyford, or the Cottesmore are streaming like a torrent in foam down the slopes from famed Ranksboro' Gorse, or Pytchley men are charging the bullfinches that raise their thorny network across the vale between Lilbourne and Crick, or the Belvoir tans are racing over the Lings from Freeby Wood to Croxton Park. Mere memory of such a moment makes the face of an old Meltonian flush and his nerves tingle with excitement. Call it wild helter-skelter confusion, fit only for boys or savages to take part in; say that it is more like the tumult of a barbaric fantasia than sport for civilized beings; hurl what epithets you will at it! But who that has once been in such a burst would not give a cycle of Cathay for those brief, rapture-laden twenty minutes over the beautiful grass lands of the shires?

Think of the throb of pulses, as a Fox breaks cover; the beating of your heart while you wait in breathless excitement for him to cross the first field; the "Tally ho! gone away!" ringing like a trumpet call, and then the answering chorus of hounds, at sound of which, like a cavalry regiment in full charge, two hundred pursuers with teeth set hard are racing for the nearest fence, and eight hundred hoofs thunder on the turf. The men who would keep in the first flight must sit down and ride then. Whatever may be of barbarity in sport dwindles to nothingness at such a moment, when noble horse and perfect horseman are inspired by common sympathies. The lightest touch of bit on mouth or a firm pressure of knees against saddle-flaps is enough to tell the sensitive animal—madly excited though he may seem—what his master asks of him as they near a blind bullfinch or treacherous oxer. To feel the heave of a clever fencer's quarters, as with neck outstretched, ears set forward, and nostrils quivering, he rises at the obstacle, is to realize why English men and women "risk neck and limb and life" in the pleasures of the chase day after day. Sweeter music man cannot hear than the rhythmic beat of hoofs on springing turf as one lands over a big fence well ahead of the charging throng and finds oneself for a moment alone with the hounds, that are skimming like sea birds across the green waves of ridge and furrow. After going for fifteen minutes at best pace, and such a pace, half the horses that were so eager at the outset have been left far behind, others are already faltering in their stride, and can only be kept going by skilful riders who know when to give them a pull, and many begin to chance their fences, crashing through quicksets or rapping the stiff top rails harder than a cautious man likes to hear. Perhaps a brief check may give breathing space to blown horses before the pack in joyous chorus hits off the scent, and men of the first flight with thinned ranks race on again for the sluggish brook "where the willow trees grow." The keen east wind bites shrewdly, and the muddy stream looks coldly grey, as if frost had already begun to grip it, but he who holds the lead at such a moment would be faint-hearted indeed if he turned aside for fear of an icy bath. Some of us don't like water, and are not ashamed to confess as much; but craven thoughts seldom come to those who can ride straight as hounds run over Leicestershire meadows when the Quorn are close upon a sinking Fox. With the excitement of rivalry thrilling every fibre of their frames, they take firmer hold of the reins, sit tighter in their saddles, give one touch of the spur, and go at it. A horse here and there may refuse; one, losing heart too late to save himself, will perhaps plunge into mid-stream, sending the spray like a fountain upward, as for a moment he disappears. The others, landing safe on the rotten banks, cast but a glance behind to see that he is not in danger of drowning, and then gallop on to where the hounds clamour fiercely over a fallen victim, and the shrill "Whoo! whoop!" tells that the brilliant burst is at an end. Such runs on a breast-high scent do not come every day, but the quickness of a huntsman in the shires will often make things merry when in slower countries hounds would be walking their Fox to death. To hunt six days a week from Melton or Market Harborough, therefore, one need be well mounted and have two horses in the field every day. With a stud of six or eight hunters one may get along very well if none of them go amiss. With more a bold horseman ought to see all the best of the fun, but with less a man had better be content to take his sport amid less stirring scenes, and not aspire to distinction in this region of fashionable fox-hunting.

FOX-HUNTING OUTSIDE THE SHIRES.

By H. H. S. PEARSE.

IN England there are a hundred hunts or more that can only be described as unfashionable in a relative sense. The habitual followers of every pack among them may be numbered by scores, or in many cases by hundreds, and as each affords sport enough for the amusement of lords and squires and ladies of high degree, they lack not the attraction that a certain air of fashion gives, though none of them can pretend to vie with the more distinguished shires. For some other packs, however, not so much can be said. Their followers make no pretence to being the arbiters of fashion in any sense. On the contrary, they treat conventionality in matters of costume with a freedom that would have made Beau Brummel shudder, and have shocked the artistic susceptibilities of a Hammond, a Tautz, or a Bartley in their most fastidious days. In such countries a man's claims to consideration are not estimated by the correct cut of a coat, the faultless fit of breeches, the exact height of a polished boot, or the way in which a bow is tied above the tops. The tailor does not make the man there, nor does anybody care one jot about the colour of his neighbour's cords, whether they be snowy white or dingy as a London fog. One point they are nearly all tenacious of, however, is that every prominent member of the hunt, unless he happen to be a parson, should appear in "pink" as a sign of respect to the Master. But it need not be the pink of perfection, and no matter how much the coat may be empurpled by weather stains, if in the original colour it conformed to regulation as by custom established. The least fastidious man in other details of personal adornment is often the most scrupulous in his observance of the respect due to the M. F. H. in this particular, and intolerant of any departure from it. Farmers, who still form a large proportion of fox-hunters in unfashionable countries, are not supposed to dress for the part. They may wear whatsoever taste dictates, though none of them would care to appear in the colour that is accepted as a badge of hunt-membership. The Meltonian is less punctilious about the sportsmanlike correctness of his costume than its style and finish, while the "provincial" exactly reverses these standards. In his opinion a man who, though entitled to wear the hunt buttons, comes out in a stable jacket, would not show a proper sense of the fitness of things; but whether the scarlet coat be old or new, the breeches immaculate, or the boots wrinkled like a Magyar's, he might not take the trouble to note. Costume, by whatever canons regulated, is, however, an untrustworthy indication of a sportsman's characteristics. Fox-hunters, like the horses they ride, go well in all shapes, and it is never safe to generalize, as Mr. Bromley Davenport did when he "held the swell provincial lower than the Melton muff." The wider sympathies and more varied experiences of Whyte Melville led him to say of the many

HUNTING OUTSIDE THE SHIRES.

countries in which he had hunted that "each has its own claim to distinction; some have collars, all have sport." Happy the man who can take his fill of the pleasure that every phase of fox-hunting affords, without discounting it by comparison with some other that comes nearer to his ideal of perfect bliss. Colonel Anstruther Thomson, who in his day was unrivalled as a rider to hounds over any kind of country, will, if I mistake not, say that he has seen quite as good sport with a rough moorland pack, the Master of which never put on a scarlet coat, as with the well-bred Pytchley, when Tom Firr and Dick Roake whipped in to him. That would be the opinion also of many a good man who has followed Jack Parker's trencher-fed hounds over the wild hills of Kirby Moorside, or heard the Blencathra chorus echoing among the fells of Cumberland. One need not, however, go so far a-field or select the very antithesis of Leicestershire sport in order to illustrate the charms that belong to unfashionable fox-hunting.

Between such extremes as the Badminton, Warwickshire, Atherstone, Bramham Moor, Grafton, and Cheshire, which even a Meltonian would not despise, and the obscure hunts of some woodland districts wherein vaulting ambition finds little scope, are many varieties of hunting country that offer attractions sufficient for all modest requirements. Their fixtures are not difficult to get at, and anybody who cares for a day with sportsmen who take more delight in the work of hounds than in steeple-chasing need not journey very far from London. He had better, however, select a country in which pheasant preserves are not plentiful, or he may only experience the disappointment of a blank day. At the trysting place he will find probably not more than fifty followers assembled, but nearly all of them are sure to be keen fox-hunters, from the veteran whose grey hairs and bent shoulders tell of three score and ten winters past, to the boy whose pony is rejoicing in its emancipation from the leading strings. A dozen farmers, young and old, who know every cover, and nearly every Fox that haunts it, are there in homely garb. The few ladies present do not affect novelties in habit skirts or eccentricities in head gear, and the horses they ride are more distinguished for cleverness than good looks. The Master, whose appearance is welcomed by all with courteous salutations, wears a hunting-cap as the outward sign of authority, instead of the silk hat which so many of his brethren have adopted since Leicestershire set that fashion. The huntsman, a weather-beaten veteran, whose wiry frame and keen face bespeak untiring energy in pursuit, has his hounds in perfect condition, and though some among them are of a type that would not find favour with judges at a Peterborough Show, they all look as if no day would be too long for them. At a signal from the Master old Jim trots away to a ride through the big woods, where his hounds, spreading right and left, are soon hard at work among the bracken and brambles of tangled undergrowth. The long succession of larch plantations and oak copses, closing here and there into deep shadowy valleys where hounds are for a time completely lost to view, would break the heart of a Leicestershire huntsman, but Jim plods on patiently from end to end, relying on himself and hounds to drive the Fox out of these strongholds if they find one here. His voice or horn is only heard at long intervals, and just loud enough to keep the hounds under command. At the first note of a light tongue thrown far ahead, he pauses for a moment with every faculty on the alert, and then, as the welcome tidings are confirmed, he cheers the pack with a shrill "Hark to Falstaff, hark!" In a minute the dead leaves are shaken with a rush and a rattling chorus, and with a few clear touches on the horn he proclaims a find. Then the old man becomes a boy once more in his enthusiasm. To force a cunning old Fox through an apparently endless chain of thickets is no child's play, but Jim's eye is quick to note every sign when the hounds are at fault, and with cheer after cheer he keeps them on the line of his hunted one. For twenty minutes they stick to the woodlands, and then an inspiriting "Tally ho! gone away!" tells that we are in for a run at last.

Horsemen and horsewomen dash along the rides at headlong speed, eager for a start. Some crash through the copse to leap the low palings of its boundary fence. Others make for a stile, in anxiety to distinguish themselves while their horses have the courage that hot rivalry gives; but the more knowing ones, who want to see the finish, however far off that may be, head for a gate that commands wide views of meadow and fallow. There is in truth no great need for hurry. A field freshly ploughed has brought hounds to their noses. They falter, then check, and Jim, who will not have them hurried at a moment so critical, gives time while they fling round in a wide self-cast. He is not of "the let 'em alone" school altogether, and knows full well the difference between hare-hunting and fox-hunting, but he never meddles with the hounds until they have done their utmost. Then his casts are made with quick decision and at no laggard's pace. The Master makes no effort to restrain impetuous pursuers. Jim's uplifted hand and politely persuasive "Hold hard, gentlemen, please!" are sufficient. Before all followers have found their way out of cover a trusty hound begins to feather on the line, gives a low whimper, then throws his tongue gaily, and all, rushing together at that challenge, drive with a merry chorus towards a valley where level meadows give promise of scent. The promise is not belied, and for ten minutes the hard riders have a taste of joy that makes them envy not the fate of Leicestershire men. Master, huntsman, and followers, however, knowing this is too good to last long, keep their eyes bent on the leading hounds, who, topping a thorn fence, speed up a swelling slope towards the flinty ridges. Into a deep goyle they plunge, and then up the far side to where a ploughman has halted his team that he may watch the sport. There the hounds come to a check suddenly. Jim, by some path known only to himself, gets to them quickly. The foremost riders, finding their way barred by a ravine too big to be jumped at one stride, make the clever hunters creep down to where the further bank affords firm space for landing on. But a young horse-dealer, whose four-year-old rushes at the goyle blindly, jumps almost on the top of Jim's favourite hounds, and only escapes the torrent of well-merited abuse by rolling backward among the bushes. This is a warning to all other impetuous folk, who are content to seek more easy ways of getting over. Some, finding a practicable bank on the higher ground, scramble over it, and rush in where more experienced sportsmen would fear to tread. Jim's self-control is sorely tried when a portly gentleman proffers advice on the assumption that our Fox must have been headed by the ploughman. Some hounds apparently thought the same, for, after a fling forward, they have come back to try the goyle. Jim, meanwhile, sits grimly silent, keeping his eye on one old hound and his horn upraised ready for a blast. "Hoick forward, boys!" he shouts joyously, as Vulcan's deep note proclaims a recovered scent. The Fox, too stout-hearted to be headed from his point, has slipped along a hedgerow unseen by the ploughman, and is setting his neck straight for some earths three miles off. But that check has given him a long start, and scent is cold on flinty hills. The hounds can hardly own to it at times, but they keep driving steadily on, and a true sportsman may well take delight in every exhibition of their sagacity, as, in their rivalry, one after another takes up the thread of pursuit. Presently their pace quickens, and we must gallop fast to catch them. They crash through a narrow shaw and race up hill towards a belt of trees. But it is too late. The Fox has found shelter in an open earth, and there one may be content to leave him with hounds baying about his stronghold. Digging for a Fox is often a necessity when the pack needs blood to give it encouragement, and to such conclusion a fox-hunting run in an unfashionable country often leads; but not every sportsman can take pleasure in it.

SNIPE SHOOTING.

By J. Moray Brown.

SNIPE shooting has one great advantage; it can be enjoyed by the poor man as well as the rich. No high rents, no keepers, no army of beaters, no highly preserved ground are necessary for its enjoyment—nor could the veriest curmudgeon of a farmer make any claim for damages inflicted by Snipe. All that is necessary is wet marshy ground, and the rest must depend on the caprice of one of the most capricious of birds. The Snipe comes and goes as the season or the weather changes, or perhaps at the ruling of some still more mysterious influence. He is here to-day, gone to-morrow; now frequenting ground where you make certain of finding him at home; at other times, and under apparently most favourable circumstances, deserting it. In fact his pursuit has always that concomitant amount of uncertainty which enhances the delights of sport. Then, too, Snipe offer as a rule such difficult and sporting shots that the knocking down of two or three couples will, in the eyes of most men not satiated with bird slaughter be more appreciated than the bagging of many partridges or pheasants.

The charm, therefore, of this particular form of sport, lies in its uncertainty, its essentially wild surroundings, and the satisfaction of finding one's game, and holding one's gun straight. I may be unduly enthusiastic, but to me there is a charm in the mere splashing through a bit of snipe bog, a thrill engendered by the "sc-a-a-pe" of a Snipe, as he shapes his tortuous flight, that the whirr of "twice twenty thousand cock pheasants on wing" never awakens. I know that my game is thoroughly wild. I have looked for him in the proper place, and approached him in the right direction, and if, as I catch a glint of his white under-wings I have "straight powder"—why, I glow with pride and pleasure. Foolish, perhaps, to let a little bird of some 5 oz. in weight influence one so; but brother sportsmen will back me in my assertion, and those who know not the delights of shooting the Snipe can, at any rate, testify to his value as a dish. Though in other climes there exist many varieties of Snipe, in Great Britain only three are commonly known: the Great or Solitary Snipe, the Full Snipe and the Jack Snipe. The first of these is a very rare bird, for few sportsmen have so much as seen the Solitary Snipe, and though some are recorded as having been shot or seen every year, they are few and far between. I will therefore dismiss him from the list with the hope that when one gets up before any reader of *A Year of Sport and Natural History*, he may "hold straight," and thus earn distinction as having shot one of the rarest of British birds; and here I may add that such good fortune once fell to my lot some eighteen years ago. The place was Cove Common, Aldershot, and on that occasion I expended five cartridges, and bagged one Solitary Snipe, two Full, and two Jack Snipe.

Though in Scotland and Ireland, and, indeed, in some parts of

England, many a Snipe is shot during the months of September, October, and November, it is not till the first week in December that he can be said to be abundant anywhere in these islands; and after the first frosts he is a very different bird from what he was previously, for he is then sharper and quicker in his flight, and in better condition. By January he will be found in fair numbers; and in nearly every patch of marshy ground, every warm spring, or tiny rill we may expect to find our little long-billed friend. But beware how you search for him in some places, or your enthusiasm may place you in an awkward predicament, for the Snipe loves quaking bogs, and if you venture too far you may souse in up to your armpits in mud, weeds, and water, and find some difficulty in extricating yourself. Under such circumstances, and having to exercise due caution in advancing on his stronghold, the difficulty of making good shooting will naturally be considerably enhanced, for there is no standing still whilst the birds are driven to you, and you have to look out for two things, your safe footing in a treacherous bog and your game. Snipe frequent queer places at times, places that border so closely on civilization and traffic that one would hardly expect to find such essentially wild birds in them. I can, as I write, recall several such incongruous localities, notably one spot on the Possil Estate near the Maryhill Barracks, Glasgow, where one day, but a few years ago, I bagged twelve couples of Snipe besides a teal. The place was an old flooded coal-pit, and abutted on to manufactories, coal-pits and a railway—whilst there were many buildings almost within gun shot. Probably by now it is drained and built over, but even at the time I mention it was sufficiently unlikely looking ground. Another capital bit of snipe ground was a strip of marshy land covered at

SNIPE SHOOTING.

high tide, which lay between the River Clyde and the railway that runs between Bowline and Dumbarton Castle. It was not more than about 500 yards wide, if as much, and extended—I speak from memory—about three-quarters of a mile. Trains and river steamers were constantly passing and re-passing, and yet Snipe during August and September used to frequent this spot in considerable numbers, and many was the day's sport I had there.

In Snipe shooting many men affect an indifference as to how they work their ground, and this indifference affects their success in a very marked degree. If you walk *up-wind* you give the Snipe an advantage. At first sight this may appear an absurdity, for most birds take advantage of the wind and fly with it, or *down* wind. There are two birds, however, that do not do this—blackgame and Snipe. They *always* rise against the wind. Let the sportsman bear in mind that if he wants to get the better of Snipe—and what is woodcraft but approaching your game under the most favourable circumstances to yourself?—he must approach the bird's haunt down wind. Then, when the bird rises he will try to face the wind and give a crossing shot, which will naturally expose more of his body than if he went straight away. Besides, the bird has then little chance of indulging in those corkscrew twists which make so many otherwise good shots miss him. As to the moment when a Snipe going straight away should be fired at, opinions differ. Some hold that you should fire directly he rises, others that you should wait till he has ended his rapid twists. On this point I will offer no advice, but merely observe that, by adopting the latter plan, you may four times out of five wait too long, and allow your bird to get out of range before the trigger be touched.

Really good Snipe shooting in Great Britain is now difficult to find, and the bags of some thirty years ago or more are difficult to make. Drainage has, no doubt, had much to do with this, and the man who now kills his twenty couples of Snipe looks on the achievement as a great one ; and yet there are places where such bags, and even better ones, may still be made.

Two or three couples of Snipe make a very agreeable addition to any bag—and there are, I fancy, few men who, when it is suggested during a day's shooting that, "we will just try a field or two for Snipe," do not feel their blood warm, and experience a sensation that the probability of any number of shots at pheasants and partridges will have failed to arouse. How anxiously on such occasions we scan every little patch of rushes in a badly drained field! How cautiously we approach the spot, and then, when the Snipe rises with his strange, familiar cry and twisting flight, how pleased we feel when we artistically cut him down !

Snipe shooting has certainly great charms, and if, after a day devoted to it, you return home wet and cold, you will not return dissatisfied, for even though your bag be a modest one, you will know that you have had a difficult bird to shoot, and a very wide-awake one ; and, if you be a gourmet, you will look forward to a *bonne bouche* in the shape of Snipe trail on toast.

WILD SWAN SHOOTING.

By George Lindesay.

A COUPLE of evenings ago old Bob, the keeper, came in from one of his moorland excursions to inform me that there were wild swans on the Black Lochs, and that, therefore, we were in for severe weather. He and the swans were right, for it is now blowing a whole gale from the north-east, accompanied by a heavy snowstorm, and I know that the wild fowl will be coming in by thousands, and that before long Bob and I will be among them.

The Black Lochs are five in number, and are distant from my home a good ten miles. There is an abominably bad peat road as far as a lonely uninhabited cottage, where we occasionally sleep when shooting the outlying beats. As far as this we can drive—at any rate, we can progress on wheels—for the remaining five miles we have to tramp it through desolate swampy flats, interspersed here and there with low hillocks.

Severe weather may with safety be predicted when the swans are seen on these most unattractive-looking tarns, but when they do arrive they are never in a hurry to leave (there being plenty of feeding), and before they do go we generally manage to get two or three. It is now three years since they paid their last visit, and then we had one of the heaviest storms known in the north for many years; but the duck shooting was grand, and besides the smaller birds we got a good many geese and swans.

After blowing with unabated fury for thirty-six hours, the gale has suddenly dropped, and at an evil hour on a bitter February morning, I begin the pleasures of the day by smashing, with the assistance of a boot-jack, the ice in my bath. Breakfast is not a lengthy meal, for Bob has been putting the pony in the cart and is now waiting at the door, as keen now for sport as he was in the days of his youth, nigh forty years ago. Each of us is indulged with a cup of steaming hot coffee with a strong admixture of brandy, and, muffled from head to foot in wraps, we emerge from the warmth of the cosy parlour into the darkness of the night. Jenny, the pony, does not at all like being dragged from her warm stable at such an unearthly hour, and exhibits her disapproval by various antics. At last, however, she discovers that she has to go, and we start off along the road with such sudden speed that "Garry," the retriever, is shot out behind with a dismal yell, and has to foot it alongside until we get a pull at our steed at least a mile further on.

Ere long the grey dawn begins to cast a little light on the desolate scene. A lot of snow has fallen, but, except under the lee of certain hillocks where the drifts lie deep, there is but little on the ground, and what there is is flattened out by the force of the wind. Some three miles from home the road approaches the head of an inlet of the bay, and here we can distinctly hear the calls of the wildfowl out at sea, the whistling of the widgeon, the "quack, quack, quack" of the mallard, the wild cry of the geese, and the

WILD SWAN SHOOTING.

clear and bugle-like note of the swans. As the light increases we ever and anon catch glimpses of the fowl circling in clouds above the sea, and when we reach the cottage there is about as much light as we are likely to have so long as the dull and leaden clouds, which look full of snow, remain.

Much to her satisfaction, Jenny is made comfortable in the little stable, the guns are carefully seen to, and the creature comforts removed from the dog-cart, and, with pipes in our mouths, Bob and I start on our tramp across the moors and swamps. Even with our local knowledge of the country, great care has to be exercised, some of those bog holes being very deep, and as they have a thin layer of ice over them, the snow lies unmelted there as on the more solid ground. Not long ago one of the shepherds got in, and would most certainly have perished, had it not been for his dog, who, seeing its master's danger, started off for the cottage, where, fortunately, we happened to be spending a few days. The dog's manner told us there was something wrong, and when we showed signs of following it exhibited unmistakable delight. When we reached the spot there was nothing to be seen of the unhappy man but his head and neck: the bog had very nearly secured a victim; even his arms, which he had very wisely stretched out when he first went in, were invisible, and certainly in another ten minutes all would have been over. As it was, we managed to rescue him in a greatly exhausted condition, and with no small difficulty, from his perilous position, and the collie's delight when at last we dragged his master on to firm ground was a sight to see.

We ascend a certain low sandy hillock, which commands a fair view of the Black Lochs, and, with the glasses, proceed to inspect the water. As luck will have it, on the nearest loch, the shore of which is not two hundred yards from us, are some thirty swans and an immense number of ducks—mostly widgeon—all mixed up together, and deeply intent on making a meal off the grass and weeds with which the tarn abounds. A powerful pair of glasses enable me to scrutinize their every movement, and for nearly half an hour we watch the scene with interest. Although a noble bird, the wild swan is not nearly so graceful or majestic on the water as its tame brother; when swimming it does not push out its wings as does the latter, nor has it the same proud carriage of head and neck. As the largest of wildfowl, it is, of course, an object of ambition to every sportsman, but it is of no use for the table, and only fit for stuffing or for a present. Some folk assert that a young cygnet roasted is not bad eating, but personally I had rather go without. Of the five Black Lochs, three are too deep to form good feeding-ground for swans, even their long necks cannot reach the grasses and weeds at the bottom; the other two are much shallower, with low grassy islands and banks, but one of these is practically useless, there not being a bit of cover for a distance of at least one hundred and fifty yards all round. There remains, therefore, only the other shallow loch further on, on the surface of which there are no big birds, as we see with our glasses. Bob and I being pretty well frozen, retire from the ridge of the hillock, and as a result of the council of war then held, "Garry" is sent forward toward the loch; instantly the air is alive with the fowl, and resounding with their various cries of alarm; but although we sit very tight and close, not a single swan comes within shot, and all make off to sea.

Then we stretch our stiffened limbs, have a stiff dose of raw whisky, light our pipes, and step out for the further shallow loch without further delay (having first of all taken the precaution to put up close to the water's edge a mild sort of scarecrow). Within ten yards of its brink there is an old "peat hag," the upper surface of which extended so far that it forms an admirable shelter for the stalker, the difficulty of course being (when there were birds on the loch) to get there unseen. On the present occasion, beyond a few ducks which fly off at once, there is nothing on the moss-brown water of the tarn to interest us, and we take up our quarters underneath the overhanging turf without any elaborate precautions.

The snow begins to descend, and I suggest to Bob the propriety of his remaining with me in the comparative shelter, but he won't have it. "If, when the swans come back, they chance to settle at the other side of the loch, they'll want drivin' towards ye," he sapiently remarks, and after a frugal meal away he goes, leaving me with "Garry" to keep me company, and both the guns, but not very sanguine as to results.

More than two hours have passed since Bob's departure, and I have pretty well given up hope, when from the direction of the sea there come clear and unmistakable notes, and flying directly towards me I can see some thirty or forty swans high in air. Need I say that I endeavour to make myself microscopically small, and need I say that Bob's precaution is justified, and that the big birds descend with a mighty swish into the water at the far side of the loch, a good three hundred yards distant from my place of concealment.

For a little time they are very much on the *qui vive*, but soon they seem to lose all sense of danger, and begin to feed and preen their feathers. Then the old man develops his generalship. I am unable to see him, or make out what he is about, but a feeling of uneasiness is somehow communicated to the swans, and they begin, without being positively alarmed, to swim and drift over in my direction. The moment for action has at last come ; the birds are within thirty yards of the shore, and I dare not put off any longer. Putting in fresh cartridges, I utter a low call taught me by my preceptor, on hearing which the birds cluster together; then I let drive with both barrels, snatch up the second gun and rush forward to the water's edge. There is a scene of wild commotion, five birds are left dead or dying on the loch, and their much-alarmed companions are getting under way as quickly as may be, lashing the surface of the water with their great wings, and uttering their wild, sonorous cries. With the first barrel of my second gun I am lucky enough to bring down a bird in the act of clearing the water ; but the next shot is a longish one, and its object gets away for the time being, to be picked up, however, later on. Bob now puts in an appearance in huge delight, for seven swans to the shoulder gun constitute a decidedly red-letter day, and we proceed with "Garry's" assistance to retrieve the game. This is no easy matter, as two of them are very lively cripples, and give a lot of trouble before they are secured ; but at length a dose of big shot terminates the struggles of the last bird, and enables "Garry" to bring it ashore. There is no time to be wasted, as the short winter's day is far advanced ; the birds are hidden under the peat bog until to-morrow, and once more we start off across the swampy wastes for the cottage, where we arrive an hour after sundown, pretty well tired out, but highly pleased with our day on the Black Lochs.

HUNTING WITH BEAGLES.

By H. H. S. Pearse.

Following hounds on foot is not a sport that commends itself to middle-aged gentlemen, who are apt, perhaps, to regard it as a frivolous pursuit, fit only for boys to indulge in. Yet those who have been bitten by a passion for that form of hunting in their younger days find its fascinations too strong to be lightly shaken off as age advances. It is a diversion that can be enjoyed to the full only by those who have strong limbs, sound wind and keen enthusiasm—all attributes of youth; but I have known many men whose love for it did not wane to the end of their lives, and, at least, one could be named who, though now a good deal on the shady side of fifty and no longer able to take the lead in a fast burst, is hard to beat at feats of long endurance. He will keep jogging along at a drover's trot, up hill or down, on firm turf or through deep plough, and nothing seems to tire him. He is, to all who try conclusions with him, a perpetual reminder of the truth that age in training can hold its own against youth out of condition any day; and if asked, he would certainly say that all this he owes to the merry music of beagles. Few of us have either opportunity or inclination to attain such a state of fitness, but the slowest may share a good deal of this kind of sport if he have the determination and stamina to see it through. Nearly all authorities, I believe, agree that Beagles which cannot go faster than the best pace of a good runner are worthless. They either do not run down their game at all, or take so long about it that hunting with them becomes no more exciting than the solution of an ingenious puzzle. From this it follows that for pursuit of stout hares on rough heath lands or even on open downs, where no obstacles except hills have to be encountered, dwarf beagles are not of much account. Nevertheless, it does not do to rush to the other extreme, for a pack that differs in nothing but in name from ordinary harriers will frequently run clean away from the field, and thus rob the sport of its most characteristic charm for all but a few, who manage to cut in at the turns. This is why horsemen are so strongly objected to in beagle hunts. A slow pack, if often excited by the presence of horses galloping abreast of it, will gradually acquire the drive that fox-hunters delight in, and a consequent speed greater than the swiftest runners can rival. Beagles thus animated by jealousy may kill hares in dashing style, but without the close and patient hunting which should be their distinguishing merit. On the other hand, a huntsman who lets them alone until they fall into habits of dwelling too long on a scent, and never rouses them to keen energy, will be in danger of losing both hare and followers. Nobody cares to watch hounds of any kind puzzling out a line as if it had got into a tangle, and as if they didn't know which was the right end of it.

An ideal pack of Beagles is such as Mr. Johnson hunts in Salop —not quite big enough, perhaps, to get over rough ground with ease, but fast enough to run a stout hare down in an hour on

scent-holding Shropshire pastures—stanch and steady on the
line as a sleuth hound; never wild or flighty, yet full of life, and
doing their work to merriest music. They are handled by one
who knows when to be patient with them, and how to quicken
their zeal with a little of his own enthusiasm, should they loiter
too long on a cold scent. That, however, is not one of their
faults. Like all good Beagles, they have a tendency to try back
in their first cast when at fault, instead of flinging forward. Such
a tendency, however, the young amateur who happens to have a
pack not quite so good as Mr. Johnson's will do well to be on
his guard against, and watch closely without interfering too
soon. The worst hounds will run heel, but the best sometimes
may not carry on so far as the hare has gone forward. With an
instinct matching that of their quarry, they are always looking out
for the twists and doubles in which puss is so cunning, and often
turn too soon, thus getting on a false scent. Then, as in all
other difficulties, the huntsman had better trust to veterans and
sages of his pack, who are sure to tell him when anything is
amiss, if he has skill to read their language of signs. Sir
Marteine Lloyd's Beagles, hunting part of South Wales, and the
Royal Rock in Cheshire, are bigger than Mr. Johnson's, but
quite as good in their work, and more adapted for a rough, hilly
country. There may be better, but if so, they make no show,
either in kennel lists or at Peterborough, and one would have to
seek their hunting-grounds in some very remote corners of the
country. Undergraduates at Oxford, and merry subalterns at
Aldershot run with packs that are not Beagles, both the Christchurch and the Divisional Foot being harriers, though no more
than sixteen inches high. Field officers who have not forgotten

the athletic exercises of their youth, occasionally join the subalterns in a run over the heath-covered ridges near camp, or through the hop gardens and meadows Farnham way, while grave and reverend seniors have at times taken a turn with the Christchurch men, but one would not advise anybody who cannot go the pace, to follow either of these two packs habitually. Though hares double and always come back, if possible, to the point whence they started, one cannot always be sure of being there at the right moment, except by running the same line as hounds or near it. On lofty, open downs a man may at times stand still and watch the chase circling hither and thither at his feet, or cut in now and then as it comes towards him, but this, after all, is only playing the part of a spectator, and not of a sportsman.

The real enthusiast in this kind of hunting will always prefer a moderately level country to hills, and if fences of strong growth alternate with broad ditches or brooks, the crossing of which can only be compassed by a brilliant effort at some risk of immersion, so much the better, for then both sinew and pluck come into play, and a man who holds his own with the pack throughout may take such pride in himself as fox-hunters who ride straight are known to entertain when they hold their own in a quick thing with a fast pack. Given good condition and sound constitution, one may, in fact, find as much pleasure in hunting with Beagles on foot as in the most costly form of sport. I have always admired as the truest sportsmen those who, not being able to afford a stud of hunters, are content to follow hounds on foot. They, at any rate, show that, with them, love of the chase is not dependent on adventitious aids. They are not buoyed up by excitement in sympathy with high courage of noble steeds, and if rivalry enters into their minds, it is of a sort that springs from no meaner motive than to do the best they can with the powers nature has given them. A well-ordered gathering with "Foot Beagles" might serve as a model for some fashionable fox-hunters to follow. One rarely sees a man out of his proper place, and sport is conducted according to methods approved by all authorities who have written learnedly on the noble science. Beagles attract a good many fair followers, and though it cannot be said that a girl is at her best when striding along in short skirts, knickerbockers, and field boots after a pack in full cry, there are some who can do this well without sacrifice of feminine grace. The majority of them, however, will prefer to leave such exercises to men, and content themselves with watching the chase from a distance that lends enchantment to a sport in which, of necessity, rough manliness finds more scope than refinement. Without being ridiculously sentimental a woman may well object to take active part in the final scenes when a timid hare is being pulled down. We know nothing, in these days, of such a system as Sir Roger de Coverley practised with his stop hounds after age had compelled him to give up his Beagles. The *Spectator* tells us of his concern "on account of the poor hare, that was now quite spent and almost within reach of her enemies ; when the huntsman, getting forward, threw down his pole before the dogs. They were now within eight yards of that game, which they had been pursuing for almost as many hours ; yet on the signal before mentioned they all made a sudden stand, and though they continued opening as much as before, durst not once attempt to pass the pole." No wonder that he was highly pleased with such a proof of discipline in the pack, and with the good nature of the knight who could not find it in his heart to kill a creature that had given him so much diversion. A modern hunt with Beagles is not quite like that, but it is, at any rate, an exhibition of animal sagacity trained to perfection, and of stout endurance on the part of human pursuers. If any doubt this, let them take their place in the field when a pack of these miniature hounds is drawing for its game, and make up their minds to follow as far as they can when the game is on foot. At first the hare goes off on a wide curve, as if heading straight for some distant coverts, and then beginning to despise the speed of her puny pursuers, she squats, with ears erect, listening for their approach. Presently, as their chorus rolls towards her, she starts again, and it seems scarcely an effort for her to distance them. But, following every turn of the scent, they press forward eagerly, and stout runners who are yet fresh enough to top the thorn fences in their stride have as much as they can do to keep up with the pack. A check, while the skein of some intricate doubles is being puzzled out, gives slower pursuers a chance of coming up, and then the merry music begins again. So, for nearly an hour, the chase goes on in circles that slowly narrow. At last the hare doubles as if in despair of being able to shake off her enemies. She is reduced to her last shifts, and unless some accident befriends her the end is inevitable. So long as scent holds the Beagles will not leave it willingly, but often in spite of their determination, the hare escapes by speed or cunning, and then some of the best sportsmen do not feel great disappointment, for, like Sir Roger de Coverley, they are loth to see a creature killed that has given them so much diversion.

FEBRUARY.

THE WILD GOOSE.

By George Lindesay.

In consequence of its size and the comparative infrequency of its occurrence, the Wild Swan is doubtless looked upon by the wild fowler as the most valuable prize he can secure, but the Wild Goose is not only a more difficult bird to get at, he is also, when in condition, an excellent addition to the table, which is more than can be said of the swan. It would be hard to name a bird more thoroughly able to take care of itself, and whose senses of hearing, seeing and smelling are more acute than the Wild Goose.

Unless at night, in a gale of wind, or in very thick weather, they fly at a height quite beyond the range of an ordinary gun. Before settling down, whether it be on land or water, they invariably inspect the neighbourhood for any sign of an enemy, and when they finally conclude that all is right, a sentinel is at once told off for duty, who, while his companions are feeding, keeps a remarkably sharp look out in all directions. This bird can communicate an alarm silently as well as in noisy fashion, the cessation of the continuous low chuckle which he keeps up while on the watch being sufficient to cause the rest to leave off feeding at once. After a certain time another bird relieves the sentinel, who then makes up for lost time with his feeding. Not only does this apply during the day when feeding, it may be, on meadows or stubble fields, but also at night, when these wary birds retire to some sheet of water for rest. Even then there is a sleepless watch bird, whose warning cry is the signal for flight. The smallest carelessness or error on the part of the stalker, the placing of his foot upon stone or gravel, the exposure of the smallest portion of his person, an eddy of wind in the wrong direction, any one of these accidents is as fatal to the sportsman's chance, when creeping up to Wild Geese, as if he were stalking a red stag in the Highlands.

However large their numbers, and however great the apparent confusion when they rise, it is curious to note the rapidity with which the birds get into the line which is at once formed and kept with methodical accuracy and precision. There are four kinds of geese which visit our shores in more or less considerable numbers, and in some cases breed within their precincts, to which the generic term "Wild Geese" is applied. These are the common Grey Lag Goose, the Bean Goose, the White-Fronted Goose, and the Brent Goose, frequently called the Bernacle.

The Grey Lag, the largest of the group, is not a winter visitor only, for, according to Mr. St. John, he found numerous nests of this species on the Sutherlandshire lochs ; and Mr. Milner, in his "Account of the Birds of Sutherlandshire and Ross," says that he found their eggs on Lochs Shin, Assynt, and Naver in the former county. They chiefly breed along the coasts of Norway and in some parts of Sweden, and are occasionally found in winter in some of the midland counties of Ireland. Their food consists for the most part of grass, and the tender and succulent shoots of young wheat, oats, or barley. The amount of destruction that a flock of Wild

FLIGHT SHOOTING.

Geese will cause in a field of either of these kinds of grain in spring is prodigious. The Grey Goose is a very good bird for the table, but his flesh is firmer and better-flavoured when he has been enabled to procure grain, on stubble fields or elsewhere. On an alarm being given all the birds run together for a second or two before taking flight, and should the stalker have been fortunate enough to get within range, a shot at this moment will give satisfactory results. This happy position, however, is one very difficult of attainment, for, as a rule, the birds avoid carefully a near approach to anything that will afford shelter to an enemy—hedgerows, ditches, and the like.

Unless, therefore, they be found on exceptionally favourable ground, driving is to be preferred to stalking, and by this means some of these fine birds are often secured. Such is the attraction afforded by newly-sown cornfields that if these be situated in a thinly-populated district, the geese will remain for days in their neighbourhood feeding upon them, retiring for the night to some conveniently situated loch or tarn among the mountains. In the month of March, Wild Geese visit some of our Scottish counties in considerable numbers in search of food of this description, and it has fallen to my lot on more than one occasion, when in Haddington and Berwickshire, at that time of the year, to bag a good many of them. While staying at a country house in the latter county some years ago, I received a visit one night from a remarkable old character, called Sandy Johnston. Sandy was a regular Jack-of-all-trades, and amongst other accomplishments he was better acquainted with the habits of the wild fowl than anyone else in the district. Knowing my liking for a shot at the Wild Geese, he had come to tell me that for the last two days he had been watching the movements of a lot of about thirty birds which had recently arrived in the neighbourhood, and that he thought we might get a shot at them if we had any luck. An early hour the following morning, therefore, saw my cousin Tom and myself on the march, under Sandy's guidance, toward the field where Sandy told us the geese were sure to be. Six miles of hard walking brought us to a high and very thick hedge about half a mile from the spot in question, a large field of young wheat belonging to an old farmer, whose dislike to the proceedings of Wild Geese was only equalled by his love of whisky.

After a prolonged stare through a small hole in the hedge, Sandy somewhat dejectedly announced that the birds were not visible. "He had seen them there," he said, "the morning before" at an earlier hour, but his advice was that we should conceal ourselves under the hedge for a time, in the hope that they might turn up later, as, owing to the presence of a certain ditch, which bisected the field in question, the place was an exceptionally favourable one for a drive. The hedge was horribly thorny and everything was very wet, the air, moreover, was uncommonly cool, and Sandy would not hear of our smoking, therefore did the time seem long and the performance monotonous.

Very nearly an hour had elapsed, when I saw the old man cock his head, and in another second or two I could hear the wild geese high aloft screaming and trumpeting; they were coming from the direction opposite to that from which Sandy expected them, and had doubtless been indulging in a meal elsewhere. After a preliminary survey, they settled down in the field and soon began to feed ravenously under the watchful guardianship of a remarkably knowing-looking old gander, who, with head erect, did sentinel's duty.

We now received our directions from Sandy. He pointed out to us where we would find the entrance to the ditch, by crawling up which we were to get into the same field, at any rate, as the geese; while he, by a circuitous route, was to reach a small spinny on the far side of the said field, and from which he was to alarm the birds quietly. Tom and I had a most horribly wet stalk of nearly three-quarters of a mile in the ditch, which had lots of water in it and was by no means too deep for sheltering purposes, but we failed to get within 100 yards of our game. We therefore lay down some fifty yards apart and awaited events with what patience we might. Suddenly the old gander gave a peculiar cry and the birds stopped feeding and ran together. Had we been within range that would have been a grand chance indeed, but almost immediately they rose, and, as luck would have it, came right over us in a confused mass. Four of the big birds fell at once to our double discharge, a fifth got away, and two others fell dead outside the great field.

The Bean Goose is one of the most frequently occurring of its family in this country; it breeds in various parts of the North of Scotland, and for six months in the year may be found in enormous "gaggles" in Tipperary, Limerick and the midland counties of Ireland; indeed, there is hardly a bog or marsh in these districts, comparatively free from human intruders, that is not frequented by them in large numbers. Except in extreme cold, when these are sheeted in ice, their favourite feeding-grounds are inland bogs and meadows, whence in the evenings they return for rest to the mud banks of the coast. When deprived by severe and continuous frost of their usual means of subsistence, they fly restlessly to and fro in search of unfrozen pools, and under such circumstances become victims of the shore shooter in large numbers.

Very similar in his food and general habits to the last-named variety is the White-fronted Goose. He is common in Ireland, and in England is frequently met with in large flocks; but in the North of Scotland he is a rare visitor. He is found in very large numbers in Sweden and Lapland, in which countries he also breeds. This goose is a capital bird for the table.

The habits of the Brent Goose, on the other hand, differ considerably from those of the Bean Goose and Grey Lag. He is rarely found inland, and seldom approaches the shore closely, his favourite food being the marine plant, *Zostera Marina*, which grows abundantly in the creeks and shoals of the coasts of many of our English counties. The Brent is the commonest of his tribe in this country, being found frequently in such vast numbers as to blacken the water or the sandbanks on which they are resting. They arrive as early as the middle of August in Belfast Lough, and do not leave until April or May; but in the North of England and in Scotland they do not make their appearance before autumn. The night is spent at sea asleep, and at earliest dawn they repair to feed to the sand-banks and shoals, where they are extremely difficult to approach. Personally I never had the fortune to make a really big shot at Brent with the punt-gun, eight couple actually picked up being my best, I think; but Sir Frederic Hughes, of Wexford, an experienced wildfowler, on one occasion, with a single discharge of his big gun carrying two pounds of shot, bagged the astonishing number of forty-seven of these birds!

RABBIT SHOOTING.

By Oswald Crawfurd.

It was an eminent living statesman who complained that rabbits had but one fault: they were six inches too short. It does not need to be a statesman, however, to have made this discovery—it only needs to shoot at a rabbit crossing a narrow ride in covert to find that a longer animal might find it less absurdly easy to get away without touch of pellet. To make this particular snap-shot requires, in my opinion, more natural quickness of hand and eye, more skill, and more practice combined than any other kind of shot at fur or feather that I know about.

If it were not for rabbits, England as a sporting country would be but a very dull one. The little white-scutted beast is an important item and incident in every day's shooting. He may start up anywhere and everywhere; from the rushy margin of a brook when we are looking for teal, snipe, or wild duck; from the underwood when we are expecting a blackcock to rise; and, when we are looking for outlying pheasants along a rough hedge side—and don't find them—half a dozen rabbits may jump out here and there and console us for our disappointment. If we look at the constituents of the great bags that the local paper loves to chronicle we generally find that rabbits make up 75 per cent. of the game enumerated.

Rabbits are hard to hit wherever found. Even in the open they do not give at all an easy shot if the ground be rough with tussocky grass, or gorse and heather tufts. Particularly when the rabbit is pressed by a dog does he cross the narrow pathway with inconceivable rapidity, and all the shooter, thirty or forty yards from him, sees is an instantaneous grey flash, and the little animal has disappeared again into the wood beyond, a fraction of a second before the gun has been raised and the discharge rung out. The ignorant man who looks on is perfectly assured that the rabbit has escaped, but he has not if the shooter knows his work; the charge has followed him, and he lies on his back stone dead, out of sight, eighteen inches within the covert. Many men can kill five out of six of these very difficult snap-shots, which seem so impossible to the onlooker. Practice makes them perfect, and nothing but constant practice would enable a man to attain to such certainty of hand and eye as I describe.

Rabbit shooting in England is the first serious shooting that a boy gets after he has done the usual snapping at blackbirds and fieldfare; and, as the walking is mostly easy and much or little can be done at the shooter's pleasure, rabbit shooting is also what the aged sportsman can best enjoy. This fact and the abundance of rabbits and the difficulty and variety of the methods of shooting them make this sport the most popular, as it certainly is the most common, of all forms of English sport.

Of the various ways of killing rabbits the present writer loves least the great rabbit battue, when no other game is shot, and where the victims are to be counted by the hundred, and at times by the

RABBIT SHOOTING.

thousand. Such a monotonous massacre may be all very well in Australia, where the conies have become a plague; but to be able to kill such a multitude in any part of Great Britain implies an injury beforehand done of *malice prepense* to the tenant and his crops. In no part of these islands will rabbits increase and multiply abnormally. Perhaps the damp climate and its vicissitudes of cold and heat, damp and dry, prevent them; or vermin—to wit, hawks, stoats, weasels, foxes, and cats run wild—with human poachers, check their over-increase. Of late years Sir William Harcourt, with his Ground Game Act, has taken rank, as a check upon the multiplication of rabbits, with the stoat, the polecat, and the poacher. Certain it is that to be able to shoot a thousand rabbits in a day over a thousand acres of land implies their over-preservation, and over-preservation means that the normal balance of nature is not preserved, and that something, sooner or later, will go amiss.

It is only for the benefit of the quite unsporting or Cockney reader that it is necessary to observe that the rabbit spends over one hundred hours every week under-ground, and that he emerges from his burrow more by night than by day. Consequently, the night poacher sees more of him than the legitimate sportsman by daylight. As rabbits are not much abroad while the sun is up, it is necessary when a rabbit " shoot " is intended to get him to leave his burrow and lie out in the fields and hedge-rows. The common way to do this has been, till recently, for the keeper to run a ferret through the burrows at night when the rabbit is not at home. The smell of the ferret is repugnant—as well it may be—to the keen nose of the rabbit, and he will not willingly re-enter the hole while the ferret's taint still lingers. This plan has its objections: where rabbits are very thick on the ground it is difficult to deal with more than a small portion of their underground haunts ; besides which, to force rabbits to stay long out of their holes is to expose them to danger from all their illegitimate foes.

There are other ways of getting the rabbit to stay afield. A piece of tow steeped in paraffin is laid at the mouth of the burrow, and two or three days afterwards, when the rabbits have gone out to feed, the entrance is stopped with a spade. This is a common way, but it is troublesome, and the burrow is tainted for a long time. A much better method, and one quite new to the present writer, is described in the Badminton Library by Mr. Lascelles. Boys are employed to stick pegs at every burrow's mouth, each peg having affixed to it a scrap of white paper. The keeper, going round his coverts, can easily see if a single hole has been passed. He carries a small bottle of spirit of tar with him, and with a feather touches each piece of paper. This keeps the rabbits out, but would not keep them out if pushed by dogs, therefore the boys are sent to stop each earth loosely with mould.

Shooting rabbits to ferrets is a sport which every real sportsman delights in. There is a certain mystery about it, an expectancy, an uncertainty, and the chance of disappointment which enhances good luck. The muzzled ferret is put down near the hole ; we all—keeper, " guns," beaters, helpers, and even the dogs—concentrate our thoughts on the little beast, as he creeps about. Does he scent his prey, and will he go in ? If he will not, it is a sign that the rabbits are from home. He enters, and we are fixed to the spot we stand on—not a word do we speak—we hardly breathe. Presently the ground under our feet seems to shake—it is the scamper of the rabbit underground—a moment more and he will bolt. Then the noise ceases suddenly—he has heard, seen, or got wind of his enemies outside—again the scamper of feet within, and this time he does bolt—suddenly, unexpectedly, as straight and swiftly as a cricket ball from a fast bowler's hand. A shot, a second, a third, a fourth, in quick succession; both "guns" have done their hardest, and in vain. There is shouting and laughing; the terriers give tongue and give chase; and before twenty can be counted the rabbit has run the gauntlet of ferret, guns, and dogs, and is safe; presently he is underground again, a quarter of a mile away.

This is not scientific sport, so conducted, but it is the very best fun in the world, and is going on in thousands of English homesteads in this frost-bound month of February. I once read in a magazine or newspaper the sentence, "the thoroughly English sport of ferreting for rabbits," but in truth it must be the least English of all our field sports, seeing that the rabbit almost certainly is not an English animal, but was brought not many centuries ago from Spain, and the ferret is quite certainly an African polecat, and could never have been fetched from Africa till rabbits were common enough in England to make it worth while to keep this very uninteresting and most unpleasant little animal in captivity. One says captivity with intention, for though half-tamed, the ferret is never really domesticated.

SPRING SALMON FISHING.

BY GEORGE LINDESAY.

THE conditions, surroundings, and influences which cause the migratory *salmonidæ* to leave salt water and to ascend during the winter months a particular river, thus constituting what is called a spring river, have always been a fertile subject of discussion and argument among anglers.

In Norway the rivers which enter the sea on the more southerly portions of the west coast invariably fish first; those in the central districts somewhat later, and last of all the rivers of Norsk Finmarkin and of the Varanger Fjord; while the further we go to eastward along the coast of the icy sea we find the innumerable streams of Russian Lapland later and later, until we reach those which enter the White Sea, which the salmon do not ascend until the month of August.

On the other hand, as regards England and Scotland, the earliest rivers are in the extreme north of the latter country, and I take it that most anglers would name the Thurso and the Naver as perhaps the earliest of all, together with those excellent Highland streams, the Helmsdale and the Brora; while the heavy fish with which Loch Tay is found stocked at the opening of the rod fishing, on the 15th of January, undoubtedly run during December and January. Again, on the whole of the west coast of Scotland the rivers are late, with but few exceptions; and so are our English streams. It may be said that the dividing point between the early or spring rivers of Scotland and the late ones is at Loch Erribol on the north coast of Sutherlandshire. From thence westward to Cape Wrath, and thence southward along the whole of the west coast as far as the head of the Solway Firth, the rivers are late; while those entering the sea between Loch Erribol and Duncansby Head, and thence southward as far as the Tweed, are early. The temperature of the sea and of the rivers, together with local conditions of the latter, no doubt account materially for this apparently singular discrepancy, and I am inclined to think with Mr. Archibald Young, the late able Inspector of Scotch Fisheries, in this matter. As he remarks, the Scottish rivers entering the German Ocean are almost all early rivers ; they have comparatively long courses, and fall into the sea at considerable distances from their sources, after flowing for some part of their career through districts not greatly elevated and possessing a moderate climate. But the German Ocean, into which those rivers run, is a cold sea ; and in winter and early spring the river temperature is, in ordinary seasons, probably higher than that of the sea, and, therefore, salmon ascend those rivers early in the year.

On the west coast, on the other hand, the rivers falling into the Atlantic are nearly all late. They have short courses, and their fountain heads are much tilted up, as they rise in that lofty and picturesque chain of mountains which, beginning in the neighbourhood of Cape Wrath, skirts the shores of Sutherland, Ross, and Inverness for over one hundred miles, at distances varying from five to twenty miles from the sea. In winter and spring, and sometimes even in early summer, these mountains are entirely or partly covered with snow, and every partial melting of their snows brings down torrents of ice-cold water, which rush through the short channels into the sea. But the water of that sea, unlike that of the German Ocean that washes our eastern shores, is warmed by the soft influence of the Gulf Stream, and the salmon therefore prefer remaining in it until the snow-water has run off, and the milder weather of June and July has raised the temperature of the river waters, and then they begin to ascend.

The case of the Norwegian rivers is different; from the North Cape southwards, along the whole of the deeply-indented west coast to the Naze of Norway, the rivers, with but few exceptions, are very similar in character, and all enter the same ocean; it is but natural, therefore, that the more southerly ones should fish first, the salmon in each case being ready to ascend them as soon as the ice breaks up and the worst of the heavy floods consequent thereupon have passed off.

Rod-fishing on the Thurso commences on the 11th of January, but for weeks before that date the salmon have been ascending it; and should the weather be sufficiently open and the water in decent order, they may be taken with the fly at once. Such, however, is not often the case, and the angler who braves the wintry storms which sweep across the dreary Caithness Flats at that time of the year has generally got to stand a good deal of " freezing out," the pools being frequently sheeted with ice.

This famous river has a course of twenty miles only from Loch More to the sea; and although the fish ascend the feeders of that loch, the fishing is confined to it and the river below. At the beginning of the season the bulk of the salmon do not go beyond the lower reaches, and to these reaches the sport is practically confined, until the rising temperature of the water induces their further upward progress. Although in the finest of condition, the Thurso salmon in the cold weather are decidedly slow and sluggish in their movements, and do not readily come to the surface; indeed, it is then necessary, in order to induce them to rise, to use very large flies, which sink a foot or two in the water. With the exception of the four or five miles next Loch More, which are comparatively picturesque, the river runs through a series of uninteresting flats, and its general characteristics can hardly be described as sporting. So slow-flowing, indeed, are many of the casts that, in order to get the fly to "work" properly, it is well to begin at the foot and fish up, as is the custom on some of the "Dubs" on Tweed. Of this type is the famous "Cottage Linn," on which such magnificent sport has often been had; but not until the month of April or May does this capital pool fish its best. Immediately above, a stretch of shallow streamy water intervenes between it and Loch Beg, a small loch just below Loch More; the result being that vast numbers of fish congregate in the "Linn" as the season advances, unwilling, unless with a flood, to ascend the shallow stream aforesaid. From ten to fifteen fish have frequently fallen to the rod of a single angler in one day, and to the best of my knowledge twenty-one is the top score; once only did I have the luck to hit off the "Linn" under favourable circumstances—the water in fair order, and a strong wind blowing from the right quarter—and I got nine fish weighing 104 lbs., besides hooking and losing at least half a dozen others.

To anyone accustomed to fish the rapid and swift-flowing streams of Norway and of the Scottish Highlands, the Thurso is by no means a tempting-looking salmon river, a fact which struck me very forcibly the first day I ever threw a fly upon its somewhat sullen waters. It was a bitter morning in early February; many of the pools were covered with ice, and a bitter "norther" was blowing, which made it a matter of no small difficulty to keep warm. However, I was in Caithness to fish, not to sit by the fireside, so, accompanied by my gillie, I started off for the Cruive Pool.

The Cruive Pool is some forty to fifty yards in width, deep and slow-flowing, and I was glad to find it fairly clear of ice; so, putting up a four-inch fly, I proceeded to fish it over. At the fourth or fifth throw there was a dead pull, and in a very brief space I hauled ashore a long, lanky kelt, which had bit the tinsel clean through and spoilt a good fly. Another kelt, and yet a third, succeeded, and I was getting very sick of pulling out the ugly brutes, when during an extremely watery "blink" of sunshine I noticed a sharpish boil in the neighbourhood of the fly. I struck hard, and found I was at last in a clean run fish. A beauty he proved, as bright as a new shilling and in perfect condition, 16 lbs.; but his play was by no means what one would have expected from such a fish. He gave me the impression of being deadened with the cold. Happening to look at the fly that the kelt had destroyed, and which I had stuck in my cap just as it was, I found the feathers stiff and hard, the wings being a mass of ice. In fact, several times that day I had to put the fly I was using in my mouth to thaw it. My next capture was another kelt, and then what was evidently a clean fish took well under water.

Thinking that perhaps I had been too gingerly with the first, I gave this one the butt unmercifully throughout. He showed his dislike of such treatment certainly, and fought a little better, but I was not impressed. He turned the scale at 13 lbs., and was also in the pink of condition. Then came a good sporting fish, for the first thing he did was to rush straight across the river, where he endeavoured to throw himself ashore, coming back again to my feet equally quickly, whereby he got so slack that I thought he must have escaped. On winding up, however, I found him still on, and after the lapse of some ten minutes we got him out. I was fortunate enough to land three other clean fish that day, but the kelts continued to be a great nuisance throughout, and destroyed several more flies before I left off.

Before concluding this paper, I will venture to draw the attention of those interested in such matters to the results of some careful experiments made at Thurso in the year 1886 by Mr. A. Harper, of Brawl Castle, which appeared in the journal of the National Fish Culture Association in July, 1887, with reference to the temperatures of sea and river which are most favourable for salmon. The blood of a salmon is always about one degree warmer than that of the water in which the fish is moving; 33° may therefore be taken as the minimum temperature of the blood—in fresh water, at any rate. An abnormally high temperature of the river water, on the other hand, is fatal. "For years," says Mr. Harper, "I have noticed with surprise the mortality which took place among the salmon of the Thurso in the months of May, June, and July, as there was not the slightest appearance of disease about the dead fish. The deaths were most numerous in places where the fish had no protection from the sun, and in slow-flowing water and well-nigh stagnant pools; but where the stream had a rapid run, and had good shelter from solar influence, such as the rocks of Dirlot supplied, the normal state of health was maintained."

Mr. Harper then gives a table of sea and river temperatures taken by him during the months of March, April, May, June, and July, 1886, which tend to show that the ordinary sea temperature is the most healthy for salmon.

From these figures it would appear that the sea during those five months had a range of 12.8°; from 39.9° in March to 52.7° in July; while the river ranged from 33.3° in March to 66.6° in July. July was exceptionally cold that year, a greater range would, therefore, have been obtained in an ordinary one. For the month of April the means of both sea and river were nearly identical, being respectively 44° and 44.7°, and during that month it not unfrequently happens that more fish are killed with the rod on the Thurso than during the whole of the rest of the season. As the river is not then, as a rule, subject to any exceptional floods, and as there are generally more fish in it in May, it would seem that those conditions of temperature account for the angler's success. That from 44° to 48° is the best temperature for the fish is further proved by the fact that when the water stands between those two degrees, old and stale fish take the fly more freely than at any other time throughout the spring. In May, June, and July, the ascending fish find themselves in water 12° or 14° warmer than that they have felt in the sea, which quite accounts for the inferior sport obtained during these months. Again, the same temperatures of sea and river recur in September and October, when the fish once more take the fly freely, and more readily than since April.

It seems a pity that such experiments as those of Mr. Harper are not more extensively and generally conducted round our coasts; they would, doubtless, go far to solve many vexed salmon problems.

MARCH.

OUR BIRDS OF PREY.

By [Aubyn Trevor-Battye.

I. THE OWLS.

TAWNY OWL.

THERE are four kinds of owl familiarly known to the naturalist in Great Britain—the Barn Owl, known also as the White Owl, the ghost-like bird that flits at dusk by meadow and wood-side, the Tawny Owl, and those rarer owls, the Long-Eared and the Short-Eared Owls.

The Tawny Owl.—That is the name by which he is most generally known, though he is sometimes spoken of as the Brown, sometimes as the Wood Owl. No owl is better known than he to those who move about at night, because he is the only owl that hoots. This hooting—one of the most marvellous and most beautiful of bird-voices—is in many districts a sound of awe to

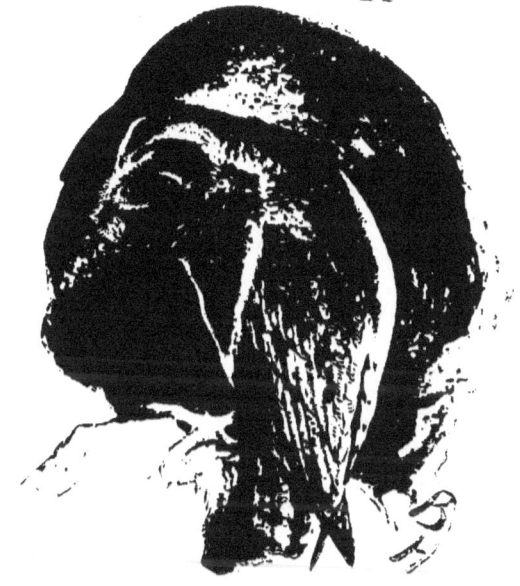

BARN OWL.

the country folk, who connect the Tawny Owl with certain superstitions, so that it is "bad luck" to kill the bird. Prejudice, sad to say, has defeated superstition in the case of the gamekeeper, who, left to his own devices by an ignorant or indifferent employer, takes this bird in his villainous pole-traps. The average gamekeeper, no doubt is above the logic of facts; we must then appeal to his master. Now, what are the facts? It is probably known to most country persons that raptorial birds eject the undigested portion of their food in the shape of pellets—" castings," as the hawking term goes. The most cursory examination of these will show that, whatever else the owls take, they do not (speaking generally) take game. The result of careful investigation by German naturalists, quoted in Yarrell (Ed. 4, vol. i.), is so instructive that it may fairly be reproduced here. It is as follows :—

	No. of pellets examined	Bats.	Rats.	Mice.	Voles.	Shrews.	Moles.	Birds.	Beetles.
Tawny Owl......	210	...	6	42	296	33	48	18a	48d
Long-eared Owl	25	...	...	6	635	...	...	2b	...
Barn Owl	706	16	3	3	93	1590	...	22c	...

(a) 1 Tree Creeper, 1 Yellow Bunting, 1 Wagtail, 15 small species undetermined.
(b) Species of Titmouse.
(c) 19 Sparrows, 1 Greenfinch, 2 Swifts.
(d) Besides a countless number of cockchafers.

Now, see. It is the vole which forms the chief food of the Tawny Owl, the destructive vole about which of late years such despairing cries have gone up from the farmers in different parts of England and Scotland ; the mole is also caught by nature's winged molecatcher, besides countless numbers of those cockchafers whose larva commits such serious ravages in the finest pastures. This is quite evidence enough to cause every game preserver, and even gamekeeper, in Great Britain to pause in the process of extermination against an innocent and most interesting bird. Beyond all doubt, the Tawny Owl is a bird that is very much the friend of the farmer and in no wise the enemy of the game preserver. If this owl takes an ill-guarded cheeper or a very young rabbit it is the exception, not the rule: it is the trifling wage his good service has fully earned.

The Tawny Owl may easily be induced, under favourable conditions, to take up its quarters near the houses of men. The writer is familiar with a pair of Tawnies which have nested for many years in one of several covered-in boxes fitted up in the trees that overhang the shrubberies in the grounds. Year after year they bring up their young, nesting sometimes in one box, sometimes in another. There are other Tawnies in the woods and parks about, but this pair are the lords of their own district, for like all birds of prey they require a large area for their hunt for food. No bird nests earlier than the Tawny and Long-eared Owls, and this pair have eggs well before the end of March, be the weather never so cold or inclement. The old cock had the misfortune to be taken in a vermin trap some years ago, and was consigned to a hamper in the stable loft with his leg in splints. Very shortly, however, he escaped and regained his freedom. But the leg was

LONG-EARED OWL.

so badly broken that the foot and tarsus dropped away. He has been on the sick-list once since then, but the loss of his foot affects him but little, and his beautiful hoot may be heard any winter night, clearly distinguishable from the voice of any neighbour of his own kind. " Kee-wick, kee-wick " cry the hungry young ones through the summer nights, the sound subsiding in a smother and a choke as the old hen stuffs the youngster's mouth with food. Long after the young are well able to provide for themselves these idle ways go on; but sooner or later there comes a day when the parents tire of their trouble, and, rounding on their shameless progeny, drive them out a-field to cater for themselves.

The Tawny displays amazing boldness in the defence of its nest. The writer has known more than one instance in which an old bird, annoyed at an inspection of her nursery quarters, has stooped with hearty good will at the head of the aggressor. That head was, fortunately, protected by a good stout hat, but the deep cuts in the head-gear made by the bird's claws bore witness to the strength of the assault.

THE BARN OWL.—As the Tawny is essentially a bird of the woods, so the Barn Owl is essentially a bird of the open fields. If the Tawny does little injury to game, the Barn Owl does still less; needless to say the gamekeeper nails it up on the kennel door all the same. A farmer of my acquaintance has allowed a pair of these birds to nest in his pigeon loft for many years, and the kindness, he considers, is on their side, not on his. So well known are the ways of this owl and its young ones that it is difficult to find anything new to say here. Its hissing, its snoring, its relays of eggs, its habit of swaying, its softness of flight, the unearthly screech with which it evokes the echoes by night and frightens the belated peasant, all these characteristics are familiar to every countryman. Round no bird has a greater wealth of legend and mystery and interest collected than round the White Owl; and this interest began away down the centuries—hundreds of years before old Gilbert White stood, watch in hand, " upon an eminence, and minuted these birds for an hour together," though that was a hundred and twenty years ago. " About once in five minutes," he says, " the one or the other of them returned to their nest."

It is interesting to notice that the Barn Owl has, like the osprey, a reversible hind toe, and this would seem to point to some different condition of existence, perhaps to the comparative scarcity of mice, and the superabundance of fish in some bygone epoch. Nowadays the Tawny is the more frequent fisher, though the Barn Owl, too, fishes in the streams and brooks at times.

THE LONG-EARED OWL.—This bird nests earlier even than the Tawny, laying its eggs by the middle of March. Though accounted a rare bird, it is commoner throughout this country than is usually supposed. It does not nest in hollow trees, but in the deserted nest of a kestrel, or in a squirrel's *drey*. No better situation can be chosen for observing the habits of the Long-Eared Owl than the fir plantations which dot the sides of the chalk hills of Surrey and Berkshire. In these clumps they nest. Later on in the year it is not unusual to find a single clump harbouring a considerable number of these birds, several families having drawn together. By looking up carefully into the trees it is always possible to see the owls, their bodies drawn up tight against the fir-trunk, till they look like bits of old wood. This owl, at nesting time, utters

70   *A YEAR OF SPORT AND NATURAL HISTORY.*

a short, sharp sound, something like the note of a frightened frog, but, of course, considerably louder.

THE SHORT-EARED OWL.—With the drainage of the fens and the cultivation of waste grounds the number of these birds that remain with us all through the year has become very greatly

SHORT-EARED OWL.

reduced. A few pairs nest still—chiefly in the east and north of this island—but it is in the autumn, when shooters are about in the fields, that the bird is most in evidence. For at that time considerable numbers come to us across the German Ocean; so that in many parts they have gained, by coincidence of arrival, the name of the Woodcock Owl. If one may compare the Tawny with the Sparrow-hawk, and the Barn Owl with the Kestrel, certainly the Short-Eared Owl may be likened to the Harrier. It beats the heaths, marches, and open places, while its singularly small head and general appearance irresistibly recall the female Hen-Harrier. Like these birds, too, it nests on the ground.

The old belief that owls were blinded by the daylight is now generally known to be a mistake; but the Short-Eared Owl is more diurnal in its habits than any of its kind with which the writer is acquainted, except the Hawk Owl.' Long before the night sets in this bird may be seen about the Broads of Norfolk, quartering the reed beds in a most interesting way. Like great moths they seem — their length of wing and lightness of body giving them a remarkable buoyancy of flight.

There are few of us who have not seen one of these poor birds rising from heather or rough scrub before the advancing line, only to be knocked over by some over-zealous shooter, resolved to let nothing off that flies. These persons, on being remonstrated with, sometimes apologize with, " Upon my honour, I thought it was a woodcock." It is a trying moment that, for one must not always say the thing that one would, nor even so much as this, " Until you have learnt one bird from another, my friend, you had better not carry a gun."

OUR BIRDS OF PREY.

By Aubyn Trevor-Battye.

II. HAWKS, BUZZARDS, KITES AND HARRIERS.

In hawking language Hawks and Falcons are distinguished into short-winged or long-winged hawks. Gamekeepers and country persons will do well to bear in mind that Falcons have notched beaks, wings pointed, and as long as their tails, and the eyes brown; and that hawks have "festooned" beaks, wings "round" and much shorter than the tail; with yellow eyes. The Falcons, from their habit of soaring and stooping upon their quarry, were in the ancient days of falconry accounted the nobler birds. The Hawks mostly course their prey through the air as a greyhound pursues a hare.

The Gos-hawk (the Goose-hawk) was never other than an uncommon bird in England, and although in the beginning of our century it was reported as breeding in the pine forests of Scotland, its presence now is confined to examples, almost invariably immature, which annually reach these shores in the spring and autumn migrations. The bird is only alluded to now with the object of entering protest against the practice of wantonly destroying such noble and interesting birds whenever they reach our shores.

In no instance is the remarkable difference in size between the

THE SPARROW-HAWK (ACCIPITER NISUS).

male and female of birds of prey so marked as in that of the Sparrow-hawk. A novice would scarcely believe that they belong to the same species. The female is three inches longer than the male, and a larger bird all round. Of a pair weighed by Mr. Gurney, the female was rather more than *double* the weight of the male. The habits of the hen Sparrow-hawk afford the one and only excuse—if excuse it can be called—for the persecution which the birds of prey undergo in this country. The cock bird is almost innocent, occasionally taking a cheeper, and that is all. The Sparrow-hawk has for the present writer an unfailing interest. Now circling the sky like an eagle in brave gyrations, now slipping adroitly through the underwood and securing a starling with unerring aim. It does everything with a certain dash and directness that separates it at once from its ally the beautiful Kestrel, though the Kestrel is a Falcon and the other is not. The older falconers reckoned this bird among the hawks " of note and worth," and an excellent little hawk it is, when trained, coming to hand quite suddenly if it comes at all. But when first taken it is so greatly given to sulks and to tumbling off the fist, that it is a terrible trial of patience, and it is safe to say that the man who can train a Sparrow-hawk can train anything.

The Buzzards are all worthless to the Falconer: they are " Varlet-hawks." No doubt many who read these lines have had, like the writer, occasional opportunities of observing our three kinds of Buzzards in this country. But those who would see them in any numbers must now, alas, go away from here to the European continent, to Siberia, or to Morocco, because, as large and noticeable birds, the hand of destruction is against them. Yet the Buzzards do not feed on game, unless on the doomed and dying young of game birds. They feed, according to their species, on frogs, snakes, beetles, grasshoppers, " wasp-grubs," and all of them on mice and moles. The "Common" Buzzard is at the present time only a little less rare in this country than its allies, the Rough Legged and the Honey Buzzard. It is, however, the only species resident with us through the year, for the second only comes in autumn as a migrant, and the third only comes to nest. No doubt the gamekeeper destroys a Buzzard whenever he can; but there need be no hesitation in asserting that the threatened extermination of the two nesting species really lies at the door of those naturalists who

THE COMMON BUZZARD (BUTEO VULGARIS).

collect and deal in skins and eggs, and whose senseless pride it is to have none but British-taken specimens in their cabinets. This enormity is only equalled by the folly of those who write to the papers and tell them where to go. Surely that is to prostitute the study of natural history. Not such was the spirit in which Linnæus, Gilbert White, and Charles Waterton showed their reverence of nature. The Common Buzzard is a fine bird on the wing, its flight remarkably resembles that of the eagle, and the writer does not mind frankly admitting that, when they are high up, he cannot tell the one from the other. This bird has, when domesticated, a curious predilection for hatching and rearing the young of other birds. There is a well-known instance on record of a female Buzzard who hatched and brought up a brood of chickens for several consecutive years.

The Honey Buzzard is placed, on good grounds, by modern systematists between the Kites and Falcons. To the Kites we are coming now. There is no need, however, to refer to any but the "common" Red Kite—the other species having occurred so very seldom in Great Britain. At the present time the Red Kite is one of our very rarest birds. There are one or two retreats in which it still abides, protected by wise sympathies. And yet, in the early part of this century, the Red Kite nested in many parts of England, Scotland, and Wales. And, just as this bird is to be seen to-day hanging about Eastern towns, so it used to be with London. It was protected to such an extent that " the Bohemian Schaschek, who visited England about 1461, after mentioning London Bridge in his journal, remarks that he had nowhere seen so great a number of Kites as there, and the statement is confirmed by Beton, who says that they were scarcely more numerous in Cairo than in London, where they remained all the year feeding on the garbage of the streets, and even of the Thames itself."

More familiar, perhaps, than the Buzzards are the Harriers. The male Hen-Harrier is known in many parts under the popular name of "the Dove-hawk." This name has probably been given to it on account of its plumage, for the difference between the feathering of the male and female is extraordinarily pronounced. The male bird resembles in general colouring many of the gulls, its back blue-grey, its under parts from light grey to white. The female—known commonly as "the Ringtail"—bears no kind of resemblance to her partner, for she is coloured brown. It is most

THE KITE (MILVUS ICTINUS).

amusing and most instructive now to read the view of the older writers on these birds, and to see the hopeless confusion into which the differences of plumage led them. What with the light plumage of the old male, the brown of the old female, the various intermediate stages of the young males, the ornithologists of that day

—Pennant, Bewick and the rest—got into a most delightful tangle. In their view the male was one species (*Cyaneus*), the female quite

THE HEN-HARRIER (CIRCUS CYANEUS).

another (*Pygassus*). It was reserved for Colonel Montagu to clear up the difficulty and upset the accepted belief. In 1805 he obtained a nest of young Hen-Harriers, kept them, and watched the young males through the stages that lead to adult dress. The Colonel's account of this may be read in his Dictionary, and is most interesting reading. He tells us how, fearing the birds might die too soon, he "forced a premature change in some of the quill and tail feathers of the others" (that is to say, he pulled

THE MARSH-HARRIER (CIRCUS ÆRUGINOSUS).

them out), and "thus compelled nature to disclose her secrets before the appointed time" (that is to say, the new feathers came, and they were grey).

That was really an achievement; and after this it seems only right that the Colonel's name should be indissolubly associated with the Harrier. And so it is—in "Montagu's" Harrier. This was, before the drainage of the fens, the commonest, perhaps, of all the Harriers. Now it is probably the most uncommon, and comes only to us (it never nested in England) on migration. In the autumn it may be seen in favourable districts quartering the marshes, and pursuing—though it probably never catches—Snipe. It has most beautiful and graceful powers of flight, and no one, I think, who has once had an opportunity of watching it thus disporting, could ever bring himself to shoot the bird.

The Marsh-Harrier, the only remaining species, is alas, like its fellow resident the Hen-Harrier, becoming rarer in England every day. The writer (although he has, of course, had his opportunities of seeing this bird when it reaches our shores in an immature condition, and has had too often occasion to lament over it as an ornament in the poulterers' shops), has never to his knowledge seen in this country a specimen that could be regarded as "resident," by which is meant a *nesting* bird; although a few pairs are reported as breeding still in certain districts. In other countries it has long been to him a familiar companion, accompanying him in his shooting rambles day after day. The Harriers differ from almost all their allies in nesting on the ground—in bog or scrub, or open heath.

When moors are reclaimed, and fens are drained, it is the ruling of nature that such birds must disappear to a very great extent. But an intelligent and fostering interest can do much on the other side. It is greatly to be desired that landlords would, for once and all, take the word of those who have made woodcraft the study of their lives, and forbid their keepers to kill these birds.

They may rest assured that their sporting interests would not suffer, and their pleasure would increase. We want our landowners and sportsmen to look upon themselves as trustees of the things of interest and beauty which nature designed for our land, and not to be content to leave to Selborne Societies and local field clubs the thankless and almost hopeless task of trying to save something of the ornithological delights before all is gone.

OUR BIRDS OF PREY.

By AUBYN TREVOR-BATTYE.

III. EAGLES, FALCONS AND OSPREY.

THE Eagle is, by the consent of all nations and of all times, the king of birds. There is probably not a people (to whom the bird is known) among whose traditions it would be possible to find a symbol more constant, a cult more old. When the Roman chose the Eagle for the sign upon his standard he took but that which had been consecrated as an object of piety by religions that had died before Rome was born. Once the "lightning-bearer" of Olympian Jove, the Eagle is still the Christian's symbol of inspiration, and is propitiated as a deity by the Indian of the Pueblo tribes.

All this is not surprising, for the Eagle has characteristics that compel respect. The dignity of his presence, the grandeur of his flight, the solitude of his surroundings seem to crown him royal, and him alone.

The very name of the Golden Eagle would have done much, no doubt, with us to give him this identity from our earliest childhood's day, even if he had never proved his title by flying away, in the story books, with babies to his nest. Not that there are no cases on record where this has really happened. I believe there really are. But the food of the Golden Eagle is for

THE GOLDEN EAGLE.

the most part not small children, but mountain hares, weakly lambs, and, alas, carrion. This propensity for eating carrion has led to the bird's wholesale destruction by means of poisoned meat. But it is very pleasant to know that now, on many deer forests, it is strictly protected. Under this wise discrimination its numbers have of late years actually increased in these islands, until at the present moment it is no doubt less rare than the Eagle we shall mention next—once, by a long way, the commoner bird.

The head and neck of the Golden Eagle are covered with feathers of a loose scale-like form and rufus colouring, and this is its only claim to the title "golden." This Eagle no longer nests in England as it once did, but in the north and west of Scotland, and in a few places in Ireland, it still breeds. It exhibits a most remarkable shrewdness in the choice of a site for its nest, which is almost always at the most absolutely inaccessible point of the mountain side. Not always, for there is a very well known instance of a poor man in County Kerry who supported his family during a summer of want by pillage of an Eagle's eyrie. He clipped the wings of the young birds so that it was a long time before they left the nest. This bird has a wide distribution. It is found in almost every European country, in Asia, on the East as far as the Himalayas, and down to the Atlas in Algeria. It is possibly this species which is in Central Asia trained to capture antelopes.

The Golden Eagle is often seen to be followed in its flight, mobbed by hooded crows, just as the smaller birds of prey are mobbed and insulted by small birds of different species.

The White-tailed Cinereous, or Sea-Eagle, is the only other British species that requires mention here. It is, as its name *Haliaëtus* implies, a bird of the coasts. Its food, none the less, is not by any means restricted to fish, for it will pick up almost anything it can get, and, like its ally, owes its diminished numbers very largely to its fondness for offal.

Mr. Wolley, in his delightful "Ootheca Wolleyana," describes nests in very small trees about four feet from the ground. The only two eyries of this species with which the writer is personally

THE OSPREY.

acquainted are in Sweden, one by an arm of the water, one by an inland lake, and both on the rock itself.

Many kinds of birds are in the habit of driving off their young as soon as they can shift for themselves—moorhens, for example, do this, as every countryman knows. The larger birds of prey can only exist under this system of isolation, and as a consequence, to a great extent, of this not an autumn passes without the appearance of eagles in this country being noticed in the papers. Most of these records come from our Eastern counties because the birds have arrived from Scandinavia and Eastern Europe. "Golden Eagles" they are often called, but they almost always prove to be immature specimens of the Sea-Eagle. Conversely the occasional immature Golden Eagle is described as the White-tailed or Sea-Eagle. The confusion arises in this way. The young Sea-Eagle has a dark brown tail, the young Golden Eagle a tail that is half white. Perhaps the easiest point of distinction to remember is this: viz. that the toes of the Sea-Eagle are "scutellated" (like a shrimp) all down the front, while in the case of the Golden Eagle, these scales or shields are reduced to three in number, situate at the distal or claw end of the toe, and the rest of the toe is "reticulate." The outer toe of the Sea-Eagle can be independently moved, and so approaches the reversible condition of this toe in the Osprey.

Few English birds—none, perhaps, but the cuckoo—have so strongly marked an identity as the Osprey; none, surely, have quite the same touch of romance. Lingering in Scotland still (it never nested in Ireland, and that in itself is strange and eclectic), it affects for its nesting-place a deserted ruin on an island in a loch, not always, but often, and in one most notable instance. In this particular case the eyrie is a structure of immense proportion, the accumulation of years and years. There seems to be very good evidence that the Osprey has really been seen to disappear under the water in its pursuit of fish, but the writer is bound to admit that he cannot answer for this from personal observation.

Anyone who has watched a kingfisher knows how extremely difficult it is to see, through the splashing and the spray raised even by this little bird, exactly what is happening. It seems to me that the kingfisher always picks up the fish from the top of the water, and yet others will stoutly maintain that he dives below it. The Osprey has less reason to dive, seeing that, while the

THE PEREGRINE.

kingfisher seizes his prey with his bill, he takes his with his feet.

The Peregrines are found pretty well all over the world. This, which is the falconers' prince of Falcons, is in the eyes of the naturalists one of the bravest birds that fly. Owing to the inaccessible situations it chooses for its nest, the Peregrine is still far from an uncommon bird in England. Few bird-nesters care to hazard a descent down the perpendicular wall of a wild chalk cliff after Peregrine's eggs. Whether the Peregrine can travel 150 miles an hour, as has been stated, is hard to say, though the experiments now being tried in Germany may help to clear up the point; but their speed of flight is of course immense, and no bird moves faster than a Falcon in its stoop. How far he can see it is impossible to say, but that his vision is good up to five miles is beyond all doubt. A phenomenon familiar enough to the falconer goes to illustrate the Peregrine's long sight. It is this. It frequently happens that when a trained Falcon is flown in a district where Peregrines have never been seen, a wild Peregrine will suddenly appear as if by magic, coming from no one can say where, to toy for a time with the trained bird. Here is a fact well established, the more difficult of explanation because it occurs at a time of the year when Peregrines are not on migration. Again, if one of a pair—no matter of which sex—be killed at nest, the survivor, in the course of the next three days, will be found to have mated again; and this has been known to happen three times over in the same instance, whether it was the male or the female that had lost its life.

The Peregrine has nested for many consecutive years in the spire of Salisbury Cathedral, and I have read somewhere an account of how a Canon went regularly to take a teal or wild duck for his dinner from the nest.

There are three British Falcons yet to be touched upon—the Merlin, the Hobby, and the Kestrel.

There is about the little Merlin a dash and go. The traveller between London and Edinburgh who keeps his eyes open is pretty certain to have more than one opportunity in any journey of seeing the Merlin stooping at larks. This bird is seldom found in closely-wooded countries, but on the moors, wolds and open districts. It is a common mistake to suppose that it always builds its nest on the ground. With us, it is true, it comparatively seldom nests in other situations, but there are exceptions. In an instance that

THE HOBBY.

came under the writer's notice, the female of a pair of Merlins being shot, the male bird next year, taking a new mate, nested in exactly the same site as before, viz. the deserted nest of a crow, well up in a Scotch fir. In Lapland the Merlin oftenest nests in trees.

The Hobby is fortunately a far commoner bird in this country than is usually supposed. But as it only arrives in late spring, when the leaf is well on the trees, and during its nesting confines itself very much to catching insects, beetles, cockchafers, moths, &c., it is comparatively little in evidence, and manages to escape the gamekeeper's wiles. In the autumn it is often seen in chase of larks, and by October it has left this land. In appearance the Hobby is a miniature Peregrine, and as a pair of these birds toy

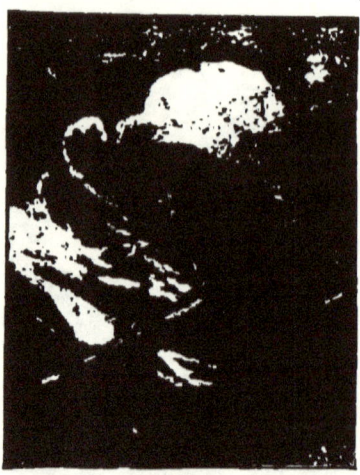

THE KESTREL.

about over the top of the oak trees in their hunt for beetles and cockchafers, the white throat and black moustaches are very noticeable, as shown by our artist.

The Kestrel, the "Windhover," is the most familiar British Falcon. The Kestrel, like the Hobby, is fond of insect-catching, though, like its near relation, the barn owl, it largely lives on mice. The Kestrel is an excellent instance of a point which I have tried to make clear before. It is this: although it is perfectly true to say that the Kestrel does not habitually feed on young game birds, it is not true to say that it *never* does so. From time to time a particular Kestrel will, no doubt, develop this taste and take to visiting the coops. Common as the Kestrel is in England, it is far more abundant in other countries. In the South of Spain, for example, these birds are quite gregarious. The writer has seen "clouds" of them wheeling round the towers of Cordova Cathedral. They are frequently seen in Devonshire perched on the weathercocks of churches or wheeling round the lofty steeple, as our artist has here represented a pair of Kestrels. There is no more charming sight to a lover of nature than the Kestrel hanging, with scarcely vibrating pinions, over down or cliff in the eye of the wind. It really does seem as if at last this bird will be protected in this country, as if the repeated efforts of accredited writers have succeeded in hammering home this truth, that in the Kestrel the farmer has a friend he can very ill afford to lose.

APRIL

BIRD NESTING.

I.—SEA-BIRDS.

THE schoolboy who goes birds'-nesting for the purpose of getting his string of eggs is quite unconscious that he is studying the elements of one of the latest of the sciences, that of Oology, which until recently has been much neglected, and, in fact, was hardly regarded as worthy of study until its philosophical bearings were pointed out by the late Charles Darwin, by Mr. Alfred Russell Wallace, and Mr. Henry Seebohm. To most persons the strange variations in the form and colour of the eggs of different birds has been merely a matter of curiosity, whereas the naturalist knows that every egg is so marked and coloured to conceal it, as far as practicable, from its enemies, and to adapt it to the circumstances and conditions under which it is laid. In no case is this more strikingly shown than in the nests and eggs of sea-birds, of which we reproduce three examples drawn from the accurate reproductions made under the care of Dr. Günther, and shown in the Natural History Museum at South Kensington. The most remarkable instance of this appears in the large case displaying the adult birds, the young, the eggs, and nest of the common Tern. Visitors looking at this example, if they are familiar with the manner in which pebbles are deposited by the action of the waves on a raised beach, will be struck with the extraordinarily accurate reproduction of the arrangement of the stones. This could hardly be done by the unassisted eye, or by memory. The secret of the success of this case is that as each

THE TERN.

stone was taken up it was numbered on the lower side, and they are all laid in the museum exactly as they were on the beach whence they were obtained. Such accurate reproduction is worthy of all praise.

Of the nest of the Tern we can say but little; in many cases there is absolutely none, the eggs, two or three in number, being laid upon the bare shingle. In colour these are brownish or greyish buff, covered with spots of dark brown, so that they look like the stones on the beach, and escape observation. But the safety of the eggs depends to a greater extent on the Tern selecting small, uninhabited islands for its nursery. Thus, the most important breeding stations in the United Kingdom are the Farn Islands, off the coast of Northumberland, where the eggs are laid amongst the shingle, and the birds nest in safety unless disturbed by the advent of a boat. The bird itself is not unfrequently termed the sea swallow, from its possessing a long forked tail, but of course it has no affinity whatever to our graceful and better known insect-catcher. The Tern is a migratory bird, arriving in England from the warm South in April and May, and wending its way to the unfrequented islands where it lays its eggs and rears its young. A flock of terns hovering in an ever-moving mass over the quiet waters of an inland loch is one of the most beautiful sights that can delight the bird lover. The mode in which they swoop down to capture the small fish on which they feed is very characteristic. It is noticeable that not only are the eggs so coloured as to be not easily discerned, but the tints of the young are mottled, and as they lie close among the stones they are hardly to be distinguished from the surrounding shingle.

Our second example, the Lesser Gull, is perhaps the commonest of all the English Gulls. It is so destructive to the eggs of game birds that it is banished as much as possible from the grouse districts of the North, and, like the Tern, has taken up its residence on many of the islands in the neighbourhood of the coast. Unlike the Terns, however, the Gulls do not confine their feeding to small fish. This species is almost omnivorous. Its food consists chiefly of fish and small crabs, the undigested shells of which are thrown up in pellets. It eats all kinds of animal substances that come in its way, and is very destructive to the eggs and young, not only of the grouse, but of the various water-fowl. It is one of the tamest of the Gull tribe, frequenting harbours, and picking up refuse thrown overboard from the ships; at other times

THE LESSER GULL.

it follows the plough, and feeds on worms and insects. The Farn Islands may be regarded as the metropolis of this species. At the nesting season the isles are white with the birds standing over their eggs, each one, as Mr. Seebohm says, standing head to wind, so that the colony looks like an army of white weathercocks. Their boldness and familiarity may be inferred from the following incident. A pair of Lesser Gulls selected for their nesting-place the middle of a sheep run. All who are familiar with the habits of these northern sheep know that they invariably follow their leader on the same track. The Gulls sitting on their nests in this run were not disturbed by the passing of the sheep, although, as the leader leapt over the sitting bird, every one of the whole flock, however numerous, followed his example.

To visitors to many of our northern islands the Puffin is one of the best known of English sea-birds. It breeds in countless thousands in appropriate situations, such as the Bass Rock in the Firth of Forth, St. Kilda, the Farn Islands, and other places. In the summer-time, when it is nesting, its numbers are almost incredible. The sea is thickly spotted with the birds, and on any alarm they seek safety in diving in preference to flight, although, in truth, their progress under water is absolute flying, the wings being used to propel them with great speed and often for a very long distance. The wings of the Puffin are remarkable, as being of that happy medium, as regards size, which enables them to be used for flight in the air and as fins when under water; the Puffin in this respect differing from the majority of diving birds that propel themselves by their feet alone. The nest of this singular bird is either deposited in a fissure of the cliffs or at some distance down a burrow in the turf, an old rabbit warren being often used for this purpose. The holes vary considerably in shape and size, and sometimes a couple of pairs will live in the same burrow. Like many other sea birds, the Puffin lays but a single egg, which is covered with faint spots, not being white like the generality of eggs laid in holes. The young bird, when hatched, remains in the nest until it is able to fly, being carefully fed by the parents, first of all with disgorged food and afterwards with small fish. The bill of the Puffin is very remarkable, and gives the idea of its being able to inflict a severe bite, but its power of doing this has been exaggerated, for during life the edges of the mandibles are somewhat fleshy. A very remarkable cir-

THE PUFFIN.

cumstance has been noticed with regard to the bill of this bird. During life it is yellow at the base, blue in the centre, and almost red towards the extremity. For some reason which is not quite clear, the bill alters in size and colour during the year. In autumn, after the young are reared, the old birds shed the horny covering of the beak in several pieces, and consequently during

O

winter the beak is smaller and much duller in colour, becoming larger and brighter in the ensuing spring. The larger size of the bill in the summer may have reference to the manner in which the birds feed their young with the herring fry and other small fish. These they not only capture while swimming under the water, but in an ingenious manner manage to arrange them transversely with their beaks, so that an old Puffin may be seen coming to its young with as many as eight small fish held transversely in its mouth, the tails hanging out on either side.

The Kittiwake is one of the commonest of English gulls, and, unfortunately for itself, one of the most persecuted. During the time that the cruel fashion prevailed of wearing the wings of gulls in ladies' hats, hundreds of thousands of these birds were destroyed in the most reckless and brutal manner. When the old birds came to lay their eggs and rear their young on the coasts, as at Clovelly, Isle of May, at Ballandra, Lundy Island, and similar places, hundreds of thousands of them were shot for the *plumassiers*. In Clovelly alone it is said that ten thousand birds were destroyed in the first fortnight of the nesting season. Mr. Saunders tells that in many cases the wings were torn off the wounded birds before they were dead, and the mangled victims tossed back into the water, and hundreds of young birds were to be seen dying of starvation in the nests, owing to the destruction of the parents. The Kittiwake, unlike the other gulls, rarely comes inland, not even searching the maritime pastures or ploughed fields along the shores for worms and larvæ like the other gulls. Its habits, however, may be studied by those who are steaming about, or who are accompanying the herring boats in the North Sea, as these birds devour large quantities of herrings and sprats, and if any broken fish are thrown over the stern they will hover in hundreds behind the boat, darting down upon the fish, usually seizing the morsel before it reaches the water. The habits of the

KITTIWAKES.

Kittiwake are most interesting to the ornithologist. It may be seen hovering over the surface of the sea in large flocks and suddenly plunging down, when the spray that it makes hides it for a moment, and it rises up with its long wings raised up above its back. Its power of flight is so great that it is perfectly at home in a gale of wind, and at times may be seen sleeping on the heavy billows with its head resting on its back. Like all gulls, it swims easily and lightly, and often alights on the water to eat its food. This consists of shell-fish, marine animals, and refuse from ships. Its note renders it very familiar, and has given rise to its name. The notes seem to resemble the syllables kitti-aa, which some persons choose to interpret as "get away," which they declare the bird says as you approach its nest. Mr. Seebohm, one of our latest authorities on ornithology, is obviously very partial to this bird, and he speaks of its interesting life during the nesting season, describing a colony not far from North Cape, Norway, where there is a stupendous range of cliffs, a thousand feet high, so crowded with nests that it might be supposed that all the Kittiwakes in the world had come there to breed. He estimates the surface of the cliffs covered with their nests at over six hundred thousand square feet, which, allowing a foot for each nest, would give a total of a half a million breeding birds. It is the custom there to fire a cannon near the colony, and, as the peal re-echoes from the cliffs, from every ledge pours forth an endless stream of birds, and before the sound has died away it is overpowered by the cries of the birds, which pervade the air so thickly in every direction as to produce the appearance of a snowstorm in a whirling wind.

When the young are able to fly these nurseries are soon deserted, and the birds spend the rest of the year wandering in search of food, and going somewhat to the south during the winter. Mr. Seebohm, like all genuine naturalists, condemns in the strongest words the slaughter of these charming birds, and the leaving the young to die by slow starvation in the nests.

GUILLEMOTS ON THE NEEDLE ROCK, LUNDY ISLAND.

The Kittiwake is readily distinguished from the other gulls by the absence of the hind toe.

The Guillemot, in spite of its being a plain, uninteresting-looking bird, is a most interesting species. It is remarkable for its habits, the localities in which it resides, its mode of progression in the air, on land, and in and under the water, and, above all, for the extraordinary variation in the colour and markings of its large single egg. Our specimens were drawn at Lundy Island, but the Guillemot is truly a circum-polar bird, to be found alike in the northern districts of Europe, Asia, and America. The greater part of the year is spent in the open sea. In the spring it frequents the rocky islands and cliffs around the coast, establishing its nurseries at Flamborough Head, Bempton, the Farns, the coast of Wales, the Bass Rock, and other innumerable localities, and the Scottish rocks and islets. There upon ledges of the rocks the hen lays her single egg, which is sometimes nearly white, blotched with black and brown, at other times of a darker colour, with richer brown marks. This pattern varies again to the most lovely deep blue or green, others being reddish and purple-brown. Some eggs are quite dark at the larger end, and beautifully mossed with brown on a creamy-coloured ground, the patterns and colours alike varying in endless variety. The egg is large in size, exceeding three inches in length, and usually of a pointed form, although the shape, as well as the size and colour, varies considerably, some eggs being found not much larger than that of a pigeon. On this single egg the old Guillemot sits nearly erect for about a month, when the young bird makes its appearance and is fed with fish by the parents. When partly fledged it is conveyed to the sea from the high eminence on which it is reared. It is generally believed that it is carried down on the back of the parent, and has been noticed to tumble off before it reached the surface of the water. In

August old and young all leave the cliffs and go out to sea together.

The wings of the Guillemot, though small, enable it to fly with tolerable facility, but its true home is on the surface of the sea and under its waters. No more interesting sight can be seen than the movements of a Guillemot, as observed through the glass sides of a tank. On the surface of the water it moves solely by the action of the feet. When underneath, the feet are not used, but the bird literally flies with great rapidity through the water, capturing with the greatest facility any live fish that may be put into the tank. As it proceeds a stream of air bubbles escape from beneath the feathers, giving the bird a most remarkable silvery appearance. Its movements under water are quick and easy, and it usually remains submerged for about half a minute.

BIRD NESTING.

By W. B. TEGETMEYER.

II.—MOOR BIRDS.

AMONG the larger birds that nest on the moors the common Peewit, or Lapwing, as it is also termed, is at once the most abundant and the most familiar. At the present season its eggs, collected in vast numbers, not only in England but from the adjacent parts of the Continent, abound in the poulterers' shops, where, after having been hard boiled, they are sold at from 3s. to 10s. a dozen, according to the abundance of the supply. In the autumn, when the old birds have collected in large flocks after the breeding season is over, they are shot by the punt gunners in thousands and hawked about the streets by the itinerant vendors.

The nest of the Lapwing is to be found in varied localities. It prefers swampy places, commons, or heaths, and breeds in rough pastures, frequently selecting some artificial hollow such as the footprint of a horse or cow for its nest. This is scantily lined with a few bents or a little dried grass, on which the eggs are laid. These are usually four in number. On the nest being approached the old bird moves quietly away, and at a little distance rises into the air, fluttering around so as to call off the attention of the intruder. When the hen has quitted the nest the eggs are difficult to discover, as they closely resemble the ground and dead leaves in colour. The newly-hatched young

THE COMMON PEEWIT, OR LAPWING.

are still more like the colour of the earth, and if a Lapwing with her brood is disturbed, she either rises at once into the air, or else reels and tumbles along the ground in the most artful manner, with the appearance of being wounded, whilst the nestlings go off in different directions, hide themselves amongst the herbage, and so closely do they assimilate to its colours that it is almost impossible for the sharpest eye to discover them.

Nothing is more beautiful than the flight of the Lapwing. Its broad wings are flapped in a regular manner, hence one of its names; and it flies round and round, changing its course and tumbling through the air, uttering the notes from which it derives its other name.

The food of the Lapwing is almost exclusively of an animal nature, insects and their grubs, worms and snails forming the chief part of its dietary. As a singular example of the variation of local habits in the way of taste, it may be stated that in Ireland, where the Lapwing is exceedingly common, its eggs are not appreciated as they are in Great Britain. They are not even collected for sale, although the bird itself is netted for the table in the autumn in enormous numbers. Than the Lapwing no bird can be more interesting to the most casual observer. As it rises from its nest, its peculiar simulation of having been wounded, and its fluttering before the very nose of a dog to draw him away from its nest or young ones, is a never-ceasing source of interest. Even old and experienced dogs who might be supposed to know better, pursue the bird which is fluttering with an apparently broken wing on the ground until at last, when some distance from the nest and young, the bird rises into the air, and with a broad flapping of its graceful wings, wheels and curls round and round as if in enjoyment of having secured the safety of its brood. This sight is to be seen in very many parts of the country, but it is only in more favoured districts that the autumn flights of the old birds are to be noticed.

The Woodcock is a migratory bird, which comes to us in great flights from the Continent about October, when numbers are killed by flying against the lanterns of the lighthouses on the English coast, between midnight and daybreak; but it is also in part an English-breeding bird, for although the majority of those

THE WOODCOCK.

that arrive in this country in October go back to the North of Europe in March, hundreds remain and lay their eggs and rear their young in suitable places in England. Even within a few years woodcocks' nests have been found at Caen Wood, Highgate, and as near to the metropolis as Streatham. The nests are merely a depression in the ground, usually in a sheltered place with a lining of dead leaves which is added during the hatching. It is not always near a marshy or wet place, where the food can be procured by the bird probing the soft ground with its bill, but is sometimes in dry situations; and it is now well known that the woodcock has the power of carrying its offspring from its nesting place to the feeding ground, the young bird being carried by being clasped between the thigh and body of the parent. The food of the old birds consists almost exclusively of common earth worms, and their appetite is enormous. One naturalist attempted to keep three woodcocks in confinement, and found it almost impossible to obtain for them a sufficient supply of the large earth worms, even by the continuous labours of one man. The custom of eating the trail of the woodcock, under which name the intestines are disguised, is one which would hardly be credited were it not well known, and if gourmands knew the conditions under which its food is sometimes obtained, it would hardly be practised.

There are a set of birds known pre-eminently as the Divers. Three species are natives of Great Britain, breeding occasionally in the north of Scotland and the adjacent islands on moors and waste places. One of the most beautiful of these is the great Northern Diver, which is so common that not less than thirty birds have been seen during one winter in Plymouth Sound; it is also abundant around the Hebrides at all seasons of the year. Except during the breeding season these Divers live at sea, obtaining their food from the shoals of herrings, sprats, and other small fish, which they catch with great ease and certainty, their progress under water being extremely rapid; it has been said they fly with the velocity of an arrow in the air. Occasionally they seek their food at great depths, having been captured in trammel nets thirty fathoms below the surface of the sea. When alarmed they dive into the water, often not rising again until they have reached a distance of half-a-mile. Although the wings are small, this beautiful bird flies well, and has been taken during migration on inland lakes and waters in

THE GREAT NORTHERN DIVER.

various parts of the kingdom. Its nest is simply made of flattened herbage and moss ; it is always placed near fresh water, often on the margin of a lake or large pond. The great Northern Diver seldom progresses far on the land, as it has no power of walking, properly so called, but pushes itself along the ground, sliding and floundering on its way. If pursued, it always prefers making its escape by diving into the water, although its flight is remarkably strong and rapid. The bird is remarkable for making the most weird and unearthly noises that can be conceived. If it is shot and wounded, it utters the most mournful cries. Young specimens that have been captured after some time become exceedingly docile, and will even come and take food from the hand, proceeding along the ground like a seal by jerks, rubbing the breast against the ground. The wings are too large for the bird to fly under water, as is the case with the Puffin, which belongs to a totally different family of aquatic birds.

BIRD NESTING.

By AUBYN TREVOR-BATTYE.

III.—TREE NESTING BIRDS.

THE best way in which to treat this subject will probably be first to take those birds that nest in the tops or branches of trees and shrubs, and then those which nest in holes, or against the trunks. It is obviously quite impossible within the limits of a short article to attempt to exhaust either group. All we can do is to point to a few of the best known instances.

The Rooks, which begin nesting early in the spring, may serve us for a start, the more particularly as they go to illustrate very well a constant general law of nature. It is this: wherever you have a class of creatures of either a predatory or semi-predatory character, whose habits are colonial or semi-colonial, there you will find a considerable proportion which never breed at all, and of these a predominating proportion of males. This opens up a very wide and most interesting subject; but here we can only point to it as a fact. Rooks are exceedingly capricious in their choice of a nesting site; but, examined carefully, their ways will be generally found to be backed by common sense. For example, they will continue nesting in a pine tree long after the tree is dead, but a dying silver poplar they will desert. They show remarkable capabilities for adapting themselves to difficult circumstances, and in this connection a book might be written on London Rooks, or, for that matter, on London Crows.

The *savoir faire* of a pair of these birds in Battersea Park at this moment is absolutely astounding. They seem particularly pleased

THE ROOK.

at having chosen an island where the ornamental water-fowl are proposing to lay. Nothing could exceed this pair of birds in boldness, and yet anyone who has lain up in the hope of shooting a pair of crows knows very well how exceedingly cautious and suspicious they are. But it is of Ravens, is it not? that the story is told by a good observer of how they could count two but not three. It was in Norfolk, and he tried to shoot them. But they would not come so long as he remained, however closely hidden. As soon as he left they returned, so he took his keeper with him and presently sent him off. But the Ravens were not to be cheated so. Then he took the under keeper also, and they three went together. Presently the two keepers left and their master remained in hiding. And now the Ravens, watching these men till they were fairly gone, thought the coast was clear and so returned to meet their doom.

But some of the shyest of birds become quite bold at nesting time. What can be more wild or shy than an autumn or a winter Missel Thrush? And yet this bird nests in the garden in the most exposed situation, and will sit close, winking a bright eye within a foot of your face as you walk by. Many birds are entirely dependent on the labours of others for their nesting places. The Long Eared Owl would fare badly were it not that it nests so early that it can make use of the old Crows' nests that are still pretty weather-tight since last year.

Think as one may, it seems impossible to generalize satisfactorily as to the reasons which prompt the different birds to nest in one way or another. Why, for example, does the Magpie, who is so supremely capable of protecting its interests by weapon and courage, go to the exceeding trouble of doming its big nest? I cannot think of any answer except a guilty conscience. And there seems to be a code of honour even among Magpies. For the Starling sometimes makes its nest in the very base of the Magpie's structure, and yet eggs and young are safe.

One never visits a heronry without being struck by the anomaly presented by these long-legged birds most uncomfortably swaying at the top of their tall trees. But the Herons are not the only birds who seem to nest where they should not. (And Herons, for the matter of that, will nest on the ground in a treeless district—in the Irish Bog of Allen, for example.)

THE HERON.

Look at the Wild Duck. Putting out of the question those foreign ducks known distinctively as " Tree Ducks," the Summer Duck of our ornamental ponds, and others, our own common Mallard constantly nests in trees. I saw a nest last year, in Sweden, for example, two-thirds of the way up in a spruce fir, on an island in the Malar. And of all the birds that one would least suspect of a habit of tree-nesting, surely the waders—Plover, Sand-pipers, &c. —are the most conspicuous. And yet the Green Sand-piper, who comes to us on passage in the spring and autumn, habitually nests even in old squirrels' dreys, in the nests of thrushes, jays and wood pigeons, at a height of some thirty-five feet from the ground.

There are exceptions to every rule; and although the lovely little Golden-crested Wren almost invariably builds its exquisite nest so that it hangs swaying underneath the end of a yew or fir-tree bough, a case may be cited here in which these birds built, two years running, against the trunk of a big chestnut tree, like any spotted Fly-catcher.

Everyone knows the Fly-catcher's nest, and everyone loves the bird that builds it. Occasionally it builds its nest in behind a bit of partly-separated bark, and the Tree-creeper does the same. And yet I have known a pair of Tree-creepers that, during many consecutive years, were at immense pains to fill up the fork between the two main trunks of a double arbor vitæ. I think they must have carried into that crack enough chips, moss and bits of pampas grass to fill at the least a gallon measure.

Marvellous is the power and persistence shown by the Wood-peckers in the drilling of the holes for their nests. But they are not very wise, for they leave all the chips lying as they fall, to tell the story of their labours. Every year they make a fresh hole. The old nests they use as sleeping places, unless they are here forestalled: for the little Nuthatch, which is quite capable of making a hole for itself, and often does so, has a way of appropriating the tunnellings of the Woodpeckers, always narrowing down the entrance to the smallest possible dimensions by the use of clay and small stones.

The Wryneck, as every schoolboy knows, is never at the pains to drill for itself, but takes possession of the most convenient hole that offers. And every boy knows, too, how valiantly it guards its

THE WOODPECKER.

eggs, and the courage needed to put a hand down into that dark hole where the creature is hissing like a snake. What can be more beautiful than the courage shown by the tiniest birds when nesting? It has been the writer's practice for many years to fix up in the garden trees movable boxes with a little hole in the side through which the birds can go to nest. All the Tit-mice are fond of these. And so, when the guests are in at luncheon they are sometimes entertained by the sight of a small box brought in, in which a Blue Tit is sitting on its eggs. The cover is removed, the bird is exposed to full view, it is passed round from one to the other, yet there it sits on, undaunted, only watching with bright eyes, and trusting things are well and chivalrous.

There are few problems more interesting, but few, it seems, more impossible to solve, than the reasons for the colourings of eggs. Take the case of hidden eggs—eggs down in the dark, in holes of hollow trees. We say they want no colour here, and that this is why the Woodpeckers and Owls lay eggs that are pure white. And this is well enough as a reason, until we are met on the one hand by such a case as that of the Redstart, whose egg is sky-blue, or, on the other, by that of the Wood-pigeon, whose egg is white. With which conundrum we will stop.

TROUT FISHING IN MOUNTAIN STREAMS.

ENGLISHMEN are reproached for talking too much of the weather; but anglers talk, think and dream of it; for all their prosperity turns upon the clouds, the wind and the rain. They hate the dry east winds of March, not because of their bronchi, but because of the Trout which will not face them, but sulk in the deeper pools while the wind blows keen and cold. At this season of the year, every man with a ten foot rod, a fly book, and a week's leisure, having access to a trout stream and possessing the soul of an angler, is thinking a great deal more of the March-brown and the yellow Trout that loves the March-brown, than of anything else in heaven or on earth. In the streams that run among mountains, the first gleams of warmth from the sun, and the first soft westerly wind, bring upon the water the gnats, midges, stone flies and other small insects that in early spring tempt the Trout, from their deeper holds beneath rocks and among the root of alder and willow, into the running streams. Then it is time to put the rod together and sally forth.

It is not given to every man to choose his day. Were it so, a not too brilliant sky, a soft, warm wind, and a certain fulness of water would probably be the choice of the mountain-stream fisher. These are most important factors in the making of a full basket. So, also, is lightness of his cast and the angler's right choice of flies ; but the most important of all is the appetite of the fish. The Trout's appetite is the unknown quantity-in fishing. If Trout took their meals regularly, fishing, in losing all its uncertainty, would lose half its pleasure. On the Trout's appetite depends the *time of the take*, and that is the great mystery of every trout river. It comes on perhaps twice in a day, and lasts sometimes but an hour, ending as abruptly as it began. While the take lasts every duffer can take Trout ; the true angler is he who can tempt the fish when they are satiated with food, or but half hungry.

In trout fishing, from the comparatively facile to the impossible is a quick transition. For instance : the stream is a deepish one, rolling along with countless ripples beneath a sheltering bank ; at your feet stretches a strip of gently sloping shingle. As you cast up and across (with consummate skill) the fly sweeps down to you again. Again, a foot or two higher this time ; and again. But no eddy breaks the dark flow of silent water.

Take the same conditions, ten minutes later. You have left the pool. In your place stands an unwelcome object—another fisherman. The preparations he makes at once excite your contempt and dislike. Obviously ignorant of the laws observed by every well-regulated angler, he stands for some minutes on the high bank and makes no attempt to conceal himself. At length, with an ugly cast of his line, he throws his fly some yards below him. A fish rises. The stranger's line is slack and he deserves to lose that fish. You sincerely hope (of course, in the interests of true art) that he will lose it! Nothing of the kind. That fish's eager appetite, not the intruder's skill, has fixed the hook in the Trout's jaws. It is a good fish, a pound and a quarter in weight—a magnificent trout for a mountain stream. The fisherman plays the Trout, handling his rod with much expenditure of muscle and no delicacy of touch, much as a waggoner might handle his whip. Oh! the fish must break. No; matter beats mind. The fish yields, the waggoner triumphs. You leave him and the pool, utterly disgusted. Why should he succeed where you have failed? Just because he happened to hit upon that particular moment when an unknown impulse prompted the fish to rise. It was the *time of the take*.

Let us imagine a mild day and a soft wind in early spring. We intend to fish one of those small "becks"—eight or ten feet in width—which tumble, helter-skelter, down the mountain-side. The lower and thickly-wooded reaches in the valley we leave behind us, and tramp on till we are some way up the mountain, and till masses of limestone rock, all huddled together in savage disorder, and tufts of coarse, grey-green grass have replaced fields and woods. High above us hovers a kestrel. With the black-faced sheep and a few mountain pipits the little falcon is the only object in sight. And for a minute we abandon ourselves to the subtle fascination of this solitude.

We are on our fishing ground. First, the great question of flies. In the early months a longish and a sober-coloured fly is advisable. As a first cast, say a March-brown or stone fly, with a small black gnat as dropper. But a grouse-hackle without wings is, perhaps, as good as any, and indeed on small streams which flow through barren and high ground, hackle-flies may, as a general rule, be relied on as the most trustworthy. Gnats,

TROUT FISHING IN MOUNTAIN STREAMS.

midges, and grass spiders are the natural food of Trout in upland streams. Their tastes are not educated up to the delicate " duns " or mayflies that delight the giants of the flat chalk stream.

And now for the secret of successful beck fishing: choose the best, and only the best, spots. Wherever the natural bed is deepened by displaced boulders or worn-out banks, there the largest fish will be lying. Avoid the shallow places—in a river the big fish are often in the shallows, in a smaller water they retreat to hiding-places behind stones or under banks. There cast, and cast lightly (no easy task !), and never fish the same place twice. The more fresh ground you cover, the more fish you will come across. If you fish a big pool in a river—unless, indeed, it be very low—a comparatively small number of its occupants will see you. But nothing escapes the watchful eye of the Trout in a brook or beck. They have a more limited horizon.

It is the habit of most local experts in the North to fish with an exceedingly long rod. It keeps the angler out of sight, but its length is not an unmixed advantage. To begin with—and this is probably the most weighty objection—a long rod spoils sport. To restore the balance of equality you require long fish; and they are not to be found in mountain streams. Next, a short rod is much pleasanter to fish with, and with it one probably makes a more accurate cast. A short rod then is best, of about ten feet in length, and a light casting line, some five or six feet long, with a couple of flies. It is certainly better to fish up-stream and not down—in the day-time at any rate. Fish which are feeding lie, naturally, with their heads up-stream on the watch for what comes down to them.

And now suppose we have fished up to a point where the beck

divides into two narrow channels, a yard or so in width. There are still, small, deep holes and strong runs in abundance. They hold Trout, but are less likely places in the spring than in the autumn. Then the fish run up as far as they can for the purpose of spawning, and before the season ends the best fish, and those in best condition, are often to be found there.

But now we must be thinking of getting home. We have still several hours in which to fish, and we can still hope for sport, fishing with a longer line, down the places that were fished up in the morning. It is no easy matter to cast with a line twice the length of your rod, but it is essential to success that the cast fall lightly and accurately. When the stream is very narrow, a single fly will be found best. It is an unpleasant, and yet a not unfrequent experience, to catch a fish on the tail-fly and the bank on the dropper.

As the stream widens, we add our second fly. Just in front the beck runs in a straight and comparatively wide course. But there to one side lies a big stone, and the water behind looks a foot or two in depth. Standing ten or twelve yards off, we drop our flies over the stone. As they touch the water—a flash! The line tightens to a "sweet resistance," and we are in a Trout of three-quarters of a pound—or, as hope tells us, a good pound weight. He rests an instant in meditation; then follows the struggle. Keeping the rod bent all the time, we let him run. In four or five minutes he will be getting tired. And now, gently steering him out of the current to the calm shallows at the side, we slip the net, if we have one, under his tail, as he turns. He is our best fish, and adds a pleasing weight to our fast filling basket.

We make our way home, lighted still by the setting sun. On either side the sheep are bleating to the growing cold of evening, as they make their way up the mountain-side for dryness and shelter, as is their wont, while the shadows lengthen, and the kine cross their path coming down towards the valley and the homestead and the byres. The signs of coming night warn us that our day's sport is ended.

MAY.

THAMES TROUT FISHING.

By E. T. Sachs.

SINCE the salmon, beaten back by the foul matter which meets him already some distance out to sea, no longer exhibits his handsome proportions in the Thames, the Trout is the undoubted king of the river. Scientists very properly decline to recognize in the Thames Trout anything but *salmo fario;* but anglers, accustomed to detect minute differences in shape and coloration, will speak of a fish as being a typical Thames Trout. It being an undoubted fact that fish partake largely of the particular locality which they frequent, small wonder need there be that the Thames Trout, living always where the stream runs swift and clear, and often amidst the swirl and turmoil of the waters that, having swept with irresistible rush over a weir, buffet awhile with the rocks below, should possess the bright silvery side which is one of the tokens by which the angler knows him. Small wonder, again, need there be, seeing that he lives in a well-stocked larder in which innumerable bleak, dace, and gudgeon jostle one another, that he should also be remarkable for an astonishing depth of body, which makes him, for his length, the heaviest Trout in the world. In saying which I do not except those wonderful fish of

New Zealand, the product of the present age, which have thriven so amazingly fast because of the stock of food, the compound interest of centuries, waiting for them to eat it. The physical perfection attained by the Trout of the Thames and New Zealand is attributable to precisely the same causes.

The Thames Trout is a spoil which no one hesitates to make his own. Anglers picture him as one holding the fort, not lowering behind weeds, like the pike, but in the full rush of water where all may behold him, and whence he issues with a dash that makes known the presence of someone of importance. The angler knows, too, that even when that rare moment arrives when the Trout unwarily seizes the barbed lure, he is far from being certain of his prey. On his hook he has one possessed of both craft and power, and until the last gasp both of these will be utilized. For the supreme moment of this struggle between man and fish, men —sane men, able-bodied and with means to do other things, and the intellectuality to enjoy them—will devote hours, days and weeks to the pursuit of the Thames Trout, knowing full well that the proportion of blank days to those bringing prizes must be sadly disproportionate, judged by the standard of other phases of trout fishing. Fishing for Thames Trout is certainly the nearest approach to piscatorial gambling that can be imagined, for, spirit you never so wisely or so well, if the Trout be not in the humour your labour is all in vain.

The accepted modes of fishing for Thames Trout are two-fold, viz., live-baiting and spinning. Once upon a time there was a strong section which affected to think that the live-baiter was but little removed from a poacher, and nothing at all from a mere fish butcher. Whether the falsity of their position has made itself manifest, or whether they realized that they were preaching to the winds, is not recorded; the fact remains that this section has not been heard from of late. Live-baiting *par excellence* for Trout is conducted with a bleak. The bleak is the Trout's most accustomed food in the Thames; professional fishermen (who are perhaps in the Trout's confidence) going so far as to assert that he prefers the bleak to any other fish. As we humans prefer the gudgeon (fried), it is perhaps fortunate that the Trout has tastes which differ from ours, or gudgeon might be selling at whitebait prices. The bleak, it need not be said, is a very delicate fish, and will not long survive rough usage, consequently the tackle used must be light. The hooks will be of the snap order and will consist of a triangle and a lip hook—no more—mounted on gut. Any other system of hooks for live-baiting in streams is inferior, for the reason that the lip hook keeps the head of the bait up stream, in which position it will retain its life and vivacity for a length of time that will astonish the practitioner who has adopted other methods. The hooks will, of course, be mounted on gut—strong, but not salmon gut; and for a float I would as soon use a bottle cork, passed on the line by means of a slit, as anything else. Lead is not required as when pike fishing. The bleak, in a state of freedom, always hovers on the surface of the water, where it acts the part of surface scavenger. The Trout never looks for it anywhere else, and the angler's cause is by no means furthered by having his bait sunk near the bottom. A Trout is not suddenly inspired to hunger by the spectacle of a bleak passing near him. When his meal-time comes on (Oh that we anglers could but know when that is!), it is then, and not before, that the Trout goes foraging; and then is it that the angler's chance arrives. Not that he may

THAMES TROUT FISHING.

not see Trout dashing about after bleak all round his bait, and even over his line, without the least notice of his lure being taken. A Trout, be it known, having once fixed his eye upon a particular fish, follows that particular fish until he obtains it, or it escapes. It will follow its intended victim right through a shoal of its fellows without deviating to the right or to the left. Consequently, it must be the angler's constant hope that the Trout will see his particular bait first.

A Trout does not seize its prey and bear it off in the stealthy way adopted by the pike. There is a plunge, a dash and a tug, and either you are fast or you are not. If you attempt to strike in the way that would be proper in the case of the pike, the betting would be overwhelmingly against your being fast.

In spinning for Trout, either the natural bait or the artificial may be used. Professional fishermen will tell you that nothing can compare with the natural bait—a bleak—fixed on a spinning flight. In the calmer, heavy waters in which many Trout lie this may be true, as more opportunity for investigation is there presented to them; but in the whirl of weirs I do not think it matters a button. Anything that spins, provided it be not too large, will do there; the thing is to find the Trout in the humour. I have caught very wary Trout on a little golden spoon, on the Bell's Life spinner (an imitation of a minnow with a bend in its tail to do the spinning), and on the spiral spinning bait; and I believe that the reason I have not caught them on other kinds of artificial spinning bait is because I have not tried them. You will not find it necessary to strike when a Trout seizes the spinning bait; but keep a very taut line on him, and look out for stray eddies which will seize the line, and bulge it in a manner that will possibly assist the Trout to eject the hooks by removing the pressure which keeps them in position. The sooner the fish is drawn out of the strong stream the better, only it is not always that the hooked Trout will permit the angler to have anything approaching his own way.

Whether you fish with live bait or go spinning, never put the line into the water without thoroughly testing every part of it. Runs from Trout in the Thames come so rarely, that no chance of a break away must be given by failure to submit the tackle to a heavy strain before using. Reliance upon his tackle lends additional confidence to the angler and power to his arm.

The salmon fisherman may like to try his luck with the fly rod. If he does so in certain spots he may be successful, as others have been before him. The best places for the fly on the Thames are Sunbury Weir, Penton Hook, Windsor Old River, and Marlow Weir. The angler need not be told where to fish when he reaches any of those places. His fly should be on the small side, and he need not hesitate to fish the same water over and over again, with decent intervals of rest.

THE TRICKS OF POACHERS.

By H. H. S. Pearse.

OF the many wise things written by Richard Jefferies, and in the writing of which he showed how keen an observer of nature he was, none contained more truth in ten words than the sentence: "All poaching is founded on the habits of wild creatures." A plain and simple dogmatism of that kind seems so obvious when said, and yet how few of us have the gift to say it! Nobody studies animal nature more closely or patiently than the Poacher, and none knows the habits of birds and beasts better. Until he has acquired that knowledge he is a mere bungler at his craft. For my own part, I candidly own that the first inkling of all the charm which field sports have power to exercise over me came through an old Poacher, the most notorious of his class in the west country, where as schoolboys we used to sit literally at his feet and try to learn all that he could teach. I am not even ashamed to confess that in his company I have, more than once, practised the tricks of Poachers with nefarious intent. I will not say where, because it is just possible that there may be no statute of limitations for such offences against the law and the squire. Towards the close of his life our mentor seldom practised the arts he was so cunning in. Not that he had grown more scrupulous or less inclined for adventure, for a series of mild punishments, inadequate as deterrents and ridiculously out of proportion to his wholesale depredations, had made him wondrous bold. His cessation from active work was the result of a curious adventure. Bob rarely, if ever, went to work at night armed with anything more formidable than a stout blackthorn, but his reputation for dexterity with that, in a give and take bout at cudgelling, made keepers wary, until at last a notice from him saying that he wanted, and meant to have, some pheasants out of a certain preserve, was enough to ensure the absence of watchers when he called. A hot-tempered, athletic young squire could not brook this tyranny, so in answer to such a notice he met the Poacher one night in a dark lane, where, single-handed, they fought it out, first with tough "ash plants," then with fists. At the tenth round Bob confessed that he had met with more than his match, and so they came to a compact. "Yer, Squoire, that'll do, I tell 'ee. You'm a man, you be, and I don't mayn to taake no more ov your vessants." "Bob, you may come whenever you like, but I'll have nobody else, and you must never go into a cover until I've shot it." That compact was faithfully kept, and thenceforth no keeper did so much as Bob to prevent anybody but himself from poaching on that young squire's preserves. But the old fellow was never quite the same after the thrashing he got then. Pride in his own prowess was gone, and he degenerated— I grieve to say it—into a mere trapper of foxes.

At this he never had an equal that I knew of. He would lie out all night and in all weathers on the bleak moor side. He could track the lightest imprint of a fox's pad for hundreds of yards, and tell with unerring certainty in which, among many "clitters of stones"—as Dartmoor tors are named in the vernacular—a robber of hen roosts had gone to ground. To do him justice, he cared only to capture the wily old gourmands. A fox of the stout wild sort, that had gone to earth after a good run, he would let "bide" in peace, for Bob was a sportsman who held that a fox which could beat "Trelawny's hounds" deserved his life. And besides, as he said, "any vule could ketch wan o' they," if he only waited long enough with his net in the right place. When the hounds had marked the red rover in, there he must come out if there was no back door. Bob's methods needed more knowledge of woodcraft. If there were any small shingle near the earth he could tell by it whether the fox had come back in haste, or with leisurely steps bearing a burden, or light. From the signs on gravel, grass, bracken, or heather, he made up his mind how long it would be before the fox must needs sally out on another foraging expedition, and made his plans accordingly. He never resorted to clumsy stone traps or cruel steel ones, and his nets were of the simplest, without bells. A man with his quick ears did not want any music louder than the angry snarl of a fox in meshes to wake him. Some he killed on the spot so that he might show mask or pad to those who paid him, but a live fox was at times worth more than a dead one to him, and he always knew his market. The sporting farmers suspected, if they did not know all, and forbade him to come near their homesteads, but he called when they were out, and was always sure of a reward from the housewives, who regarded him as custodian of their perquisites.

The methods of Poachers in trapping, netting, or snaring game differ so much that one never knows how to meet them. Rabbits, as a rule, they leave to gipsies, who are experts in setting wire snares. A rabbit never bolts from its hole in a hurry, unless pursued by a ferret or stoat, but reconnoitres cautiously, and any recent disturbance of the ground by hand of man is enough to put him on his guard. Then he goes back and tries to go out some other way. Knowing this, the gipsies smear their hands with moist earth before setting a wire, and where they have laid a trap they rake leaves or earth over it lightly with a stick, not venturing to use a finger. If a professional Poacher stoops to rabbit-catching, in absence of higher game, he does it wholesale, with nets that match the hue of grass when hazy evening light is on it, so that when spread their fine meshes look only like a film of mist. Walking along the head of a cover in the dim twilight of summer or hazy moonlight of late autumn, when rabbits are out feeding, the Poacher sticks an iron rod of his net in the ground and then proceeds to set the snare, which may be a hundred yards long, or more, and this he accomplishes almost as fast as he can walk. He wants, however, one assistant to hold the first rod, and another with a lurcher to drive the rabbits back. When alarmed, they come helter-skelter, by tens or scores, rolling in the net they cannot see. When once a rabbit's head is well through a mesh he cannot hope to escape, in spite of all his struggles and terrified beating of feet. One smart tap behind the ear settles him. I have spoken of a lurcher; that is the Poacher's best friend when in search of rabbits, hares, pheasants, partridges, or black game. The lurcher is a consummate actor, or rather pantomimist, for he never utters a sound while at work, and his training is the greatest triumph of the Poacher's art. When, by a wave of the hand, a signal is given for him to go off

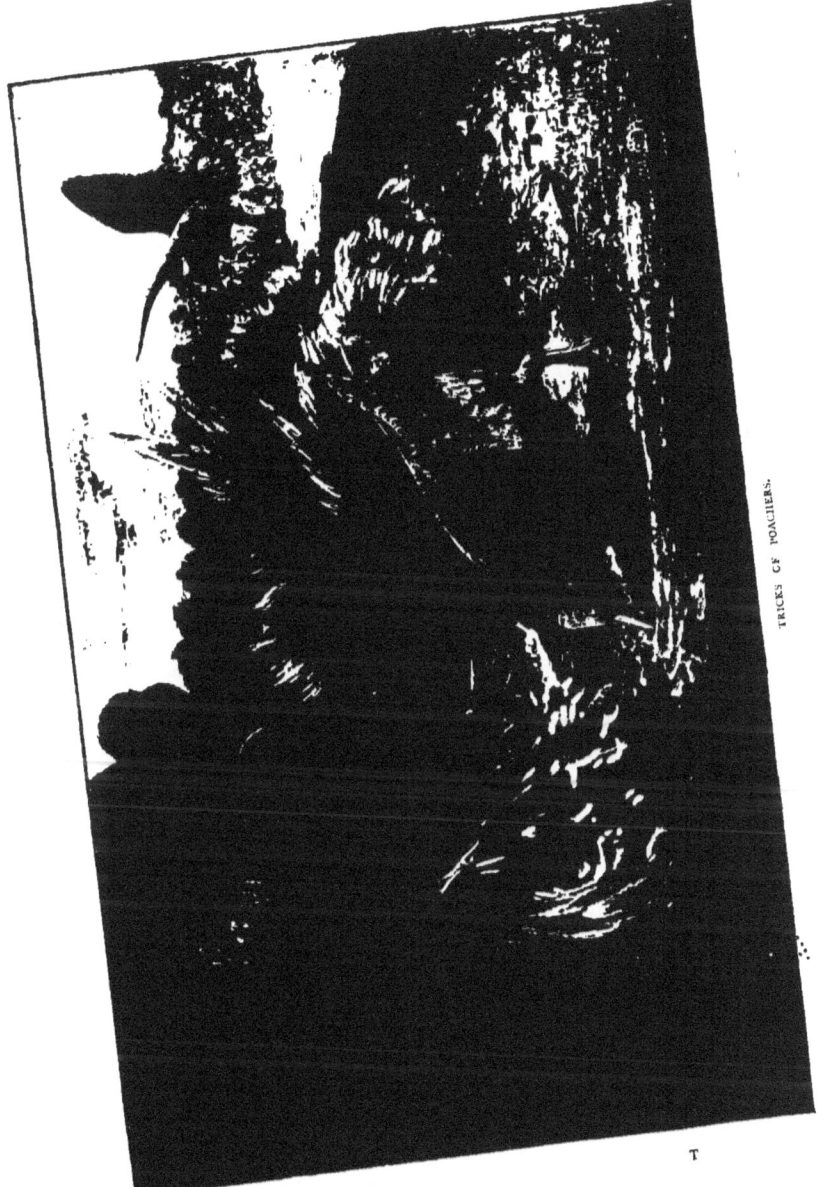

TRICKS OF POACHERS.

T

and find a hare, he quarters every yard of ground with marvellous patience. Meanwhile the Poacher sets his net in a gateway, or at a stile, or a gap, where by woodcraft he has learned that the hare will run. If all goes right, he has puss in his ample pocket before she has time to squeal. Should, however, a stranger approach, the lurcher, ever on the alert, gives timely warning, not by so much as a growl, but by going quickly away from, instead of towards his master, whom this intelligent dog has been taught not to recognize or make friends with in the presence of a stranger. I once knew a neat trick of poaching done in broad daylight. On fresh-fallen snow I came across the tracks of a hare, which seemed to have been going slowly towards its form. Following every turn and double were the footprints of a lurcher. Guided by them, I came to the empty form, and thence could follow by sight of more hasty footsteps all the windings of a chase, until they ended at a gateway, where the faint markings of a net, a few drops of blood, and a yellow stain on the snow showed clearly what had happened. From the gateway, up a narrow lane, the dog's tracks kept company with those of hob-nailed boots until all traces of them were lost amid the wheel ruts on a much-frequented highway. When a lurcher is used to bring young black game to a Poacher's gun or net on the open moor, he has a trick of loping along at a gait so closely resembling that of a shepherd's tired dog that he at times deceives the shyest old birds. Next to his lurcher, perhaps, a Poacher's best friend, though unwillingly so, is the keeper, one of the "wild creatures" on knowledge of whose habits success in poaching depends greatly. Keepers, with all their zeal and watchfulness, fall into grooves about which the fraternity of Poachers can easily learn all they want to know. Then they go to work with confidence.

In lonely copses, far from the haunts of men, and rarely visited by the keeper more than once in twenty-four hours, Poachers prefer broad daylight for their work. Amid the tangled undergrowth or long grass they can mark the runs of pheasants and hares, and learn the way they most frequently go to their feeding-grounds in the neighbouring corn-fields. Somewhere in these runs they set wire snares with a stop loop in each to ensure that whatever may be caught will not get strangled. They want the pheasants alive if possible. Another trick is to throw up a light fence of twigs and brambles from side to side of the copse, with a series of holes in it just large enough for a pheasant to pass through. In these the sensitive snare is set and the result is nearly always the same. Such a fence, however, requires time for its construction, and the work is generally done at night in readiness for operations at daybreak, when pheasants come down from their perches to feed. A few handfuls of grain artfully distributed do not come amiss, but the Poacher's favourite device is to place himself at one corner of the covert and tap two sticks together lightly, making just sound enough to drive pheasants at a run from him, without alarming them into flight. So he goes from side to side of the copse until he has reason to believe that something is caught in nearly every snare, and then home with a sack full of live game in his cart. Of numberless other tricks, more or less clever, the Poacher is master, but his greatest triumphs are of daring rather than cunning, though the latter quality always comes into play. He likes the keen, frosty air of night when clouds are drifting across the moon. Then, in company with two or three trusty accomplices bearing fowling pieces or air guns, he makes his way to well-stocked coverts, where scores of pheasants are sure to be perching on wind-stripped oak branches or under the spreading eaves of spruce firs. His lurcher is there, of course, but kept in at heels or held by a comrade who crouches behind a thicket, while the gunner creeps forward. On spruce branches, low down, pheasants may be killed, one after another, by blows from a cudgel, or even caught by the leg by one hand, while the other stops their cackling. Those on the oaks and beeches, however, perch so high that they can only be seen when in silhouette against the pale halo of veiled moonlight. But though thus out of reach, they are near enough to be brought down by a small charge, so the Poacher takes half the powder out of his cartridges, thus lessening the volume of sound when he fires. Not that he concerns himself much then about the chance of his shot being heard. Probably he is far off before any watchers have made up their minds exactly where the last report came from. Ears are deceptive when startled suddenly at night, and the Poacher takes care that nobody shall see the flash of his gun, which he discharges from behind an impenetrable screen of bushes well inside the covert, and seldom on the fringe of it. At worst, however, the risk of bringing keepers down on him for a sharp hand-to-hand tussle adds zest to the game, and in a stern grip, when wrestling for a fall, he can often show them that the tricks of Poachers are not confined to setting snares for timid or helpless wild creatures. It is impossible for a sportsman to write of Poachers without some admiration for their skill in woodcraft, though he must wish it were employed to better ends. Even the squire who suffers has a soft place in his heart for the dexterous or bold Poacher, and lets him off with a light sentence.

FISHING WITH THE DRY FLY.

By Oswald Crawfurd.

To catch trout with the wet fly in the common way is a beautiful art, but to take them with the Dry Fly is less an art than a science, or rather, it is something of both. The older method has been uncharitably described, by the followers of the new, as the "chuck and chance it" style. The fisherman throws his two, three or four flies in any water likely to hold a trout, and relies on a fish seeing and taking the lure. For aught he knows, there may not be a fish within ten yards of his cast, and he cares little whether his lure sinks below the surface or remains upon it. Not so in Dry Fly fishing. Here the angler must first discover the fish, then send the fly—the right fly—floating down the water exactly over its head. It takes a better man to do all this than to fish in the older method, a man with keener sight, for he has to guess the presence of a trout on the feed by indications which the ordinary fisherman would only suppose to be the swirls and circlets and ripples of the flowing stream. He must have more skill, too, in casting, for he must hit the water with his fly to an inch-breadth; he must also be a fair entomologist, for unless he knows most of the insects that people the river bank, and is quick to place their representative at the end of his line, he may expect no sport.

Dry Fly fishing is an invention of late years. In delicacy and difficulty it compares with common Fly fishing as that does with worm-fishing in a flooded stream. In Dry Fly fishing, the angler walks up the bank of the stream, curiously scanning its surface. He comes to the " tail of a pool." Two or three small trout rise at intervals in the troubled water, leaping bodily above the surface more in play than hunger, but the fisherman passes on. In chalk streams, where trout are fat on the good living the waters afford, fish run heavy, and these small fish are not sizeable occupants of the basket of a serious angler. He therefore disregards them, and presently comes to where the water deepens between lofty banks ; an alder stump projects from the opposite side, and his eye dwells on the swirl that the trunk and roots cause in the river's current. Just where the flowing water lines, like an elongated S, melt into the general ripple of the stream, his keen eye notes an intermittent movement in the water. A casual observer might look at it for five minutes with a field-glass, and see nothing beyond the interrupted lines of water caused by the tree trunk ; but those oily-looking undulations, and every two or three minutes those little, half-imperceptible eddyings, are caused by the movements of a heavy trout lying some six inches below the surface; his head is, of course, up stream, and he is waiting for the dun-flies and gnats, and alder and sedge-flies, that float down upon the current. While in an ordinary cast of flies the insects are as often beneath as on the surface, the object of the Dry Fly fisher is to imitate the action of the natural fly as it floats down, with its filmy wings upraised ; the imitation of the fly must be perfect, and it must be kept dry, or it will not float. Consequently the angler must whisk his line and his single fly twice or thrice through the air between every cast on the water. As soon as the fisherman has ascertained the position of a feeding trout, and guessed the manner of fly it is feeding on, he throws his imitation of it with a cast so contrived that it shall fall delicately within two or three feet up stream of the head of the fish. The fly sails down stream, but some six or eight inches to one side; the well-fed trout is too fat and lazy to move; the angler lets it float on, and only when it is well behind the fish does he flick it gently from the water; he makes two or three casts in the air to dry his fly, and again throws with better aim above the feeding fish. This time it passes within an inch of the trout's head, floating delicately with wings poised and body resting so naturally on the water that the fish is deceived; a little circlet on the water is the only sign that the trout has taken the bait. The angler waits the fraction of a second, then rather tightens the line with a firm pressure of his hand than strikes. He has a firm hold of a strong two pound trout, and before the struggle actually begins he is careful to draw the trout down stream away from the subaqueous weeds and tree roots for which the fish will instantly make. Then the fight begins, the angler always trying to draw him down stream, the fish dashing hither and thither towards the deeper pool. In three minutes, the landing net under the tired fish ends the contest.

The skilled mountain-stream angler, who has never failed to carry home his ten to twenty pounds weight of fish, wonders how, in a river full of fish, he can take but two or three tiny troutlets in a day. The reason is to be found in the abundance of insect food in these southern chalk streams. Their well-fed inhabitants, very unlike the hungry trout in a mountain stream, are fastidious about flies that are not served up to them, as it were, in the most appetizing way, and that are not almost forced into their mouths. The delicate reception of a fly by a chalk-stream fish is very different from the mad plunge, half across the stream, of a hungry mountain trout.

JUNE.

SCOTCH LOCH FISHING.

By J. W. Fogg-Elliott.

Loch fishing is not now what once it was. In some of the well-known Lochs there are not half the trout there were ten years ago. Then, any duffer could kill them; but now the bad fisherman has a very poor chance of sport. The trout have been educated, and it is the duffer who has educated them. He rises fish after fish—when there is enough wind to help him to get his flies out—pricks half of them, and perhaps catches one fish in a dozen rises. In a few out-of-the-way places, however, there still are Lochs where the trout have scarcely ever seen an artificial fly. I came across one of these last summer when climbing over a shoulder of Ben More, in Sutherland, on my way to the Gorm Loch. My gillie contemptuously called it a "peat hole," for in extent it was not more than an acre. It was connected with the large Loch by a small stream. The little pool was full of weed, but I thought I would try it; so putting on a single fly—a " Zulu "—I cast into one of the openings in the weed. Immediately a trout took the fly, and before I left I had taken five, weighing from four and a half ounces to two pounds each.

The " Zulu," used as a "top-dropper," is unquestionably the

best Loch fly in Scotland. On some days the trout will only have it when fished deep; on others they prefer it bobbing over the top of the ripples. A "Claret and Mallard" comes next, I think, in order of merit. Certainly it is the best fly on Lochs Rannoch and Luydon in Perthshire. Then the "March Brown," "Red and Teal," "Heckum Peckum" and "Green and Grouse" are good. Different Lochs have their favourite flies, but that which was best on a certain Loch one season may not be equally good the next. For instance, on Loch Awe, in Sutherland, the "Red and Teal" was by far the best fly in June, two years ago—last year, during the same month, the trout would have nothing to say to it. But perhaps this Loch is exceptional, for during June enormous rises of May-fly come on, and the trout do not rise freely to anything else. Only on one other Loch in this neighbourhood did I see any quantity of May-fly; on Loch Assynt, only three miles off, I never saw even a solitary specimen. No doubt this accounts for the trout on Loch Awe being exceptionally fine fish. On an ordinary day they will average a pound—and larger fish are not at all rare—while in Assynt, a Loch at least eighty times as large, the trout only average three to the pound.

I have found a small "Red Spinner" tied on a drawn gut very useful on calm, bright days. By casting from the shore into the rings made by the rising trout, good sport may be had. In connection with this, I once saw a rather amusing incident. At the bottom of Loch Assynt there are a number of springs, which are continually throwing up large bubbles. These from a distance appear exactly like fish feeding on the surface, and I once saw a man wade carefully out to within twenty yards of these bubbles. He then proceeded to offer them various flies, and fished most industriously for a quarter of an hour, when his gillie appeared and told him what he was fishing for.

A certain class of anglers frequently cause a good deal of unpleasantness; they arrive at an inn, where perhaps six or seven men have been staying for weeks, and expect to have a boat on one of the best Lochs next day. This can hardly be considered reasonable. It is all right when there are, say, four men and four good beats; but when three or four more men arrive, then the names of an equal number of Lochs are put into a hat, and each man draws for his beat. The names of the Loch and the angler are written down, and the men move up one each day. Anyone arriving after this is put down at the bottom of the list. This seems the fairest method.

I think trolling for large trout is the best part of Loch fishing. It is the perfection of sport, for there is nothing simple or confiding about *salmo ferox*. It will take the finest tackle you can safely use, and an accurate knowledge of the habits of the fish, before you are successful. Most authorities put *salmo ferox* down as a distinct breed, but in what way does he differ from a common trout of equal size? His colouring is certainly darker, and he lacks the red spots of the smaller trout; but this can be accounted for from the fact that he retires to the deeper part of the Loch during the day; and if you put a brightly spotted trout in a deep, dark pool, he soon loses his brilliant colour. The leopard may stick to his spots, but I am not so sure of the trout following suit. These large trout very rarely come on to the shallows, except at night to feed, and it is seldom one is taken on a fly. Yet there must be hundreds, or thousands, of them of over ten pounds weight in such lochs as Assynt, Erricht and Awe. The evening is

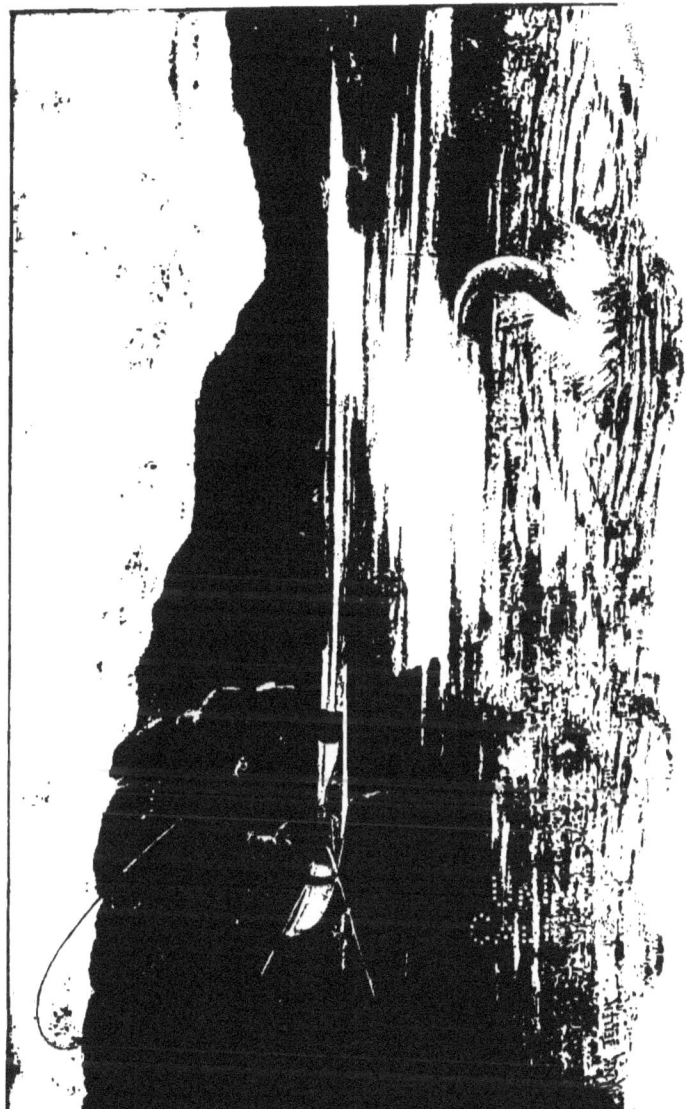

LOCH FISHING.

generally the best time for trolling, and I know nothing pleasanter than rowing quietly along on a still June night. You can hear the hum of voices from the inn, more than a mile away, and the regular beat of the oars makes you drowsy. Suddenly the rod you are holding is nearly jerked from your hand and the line rushes off the reel. To pass the other rod to the boatman to wind up does not take a second, and the fun begins. Eighty yards of line and an ounce of lead won't stop the fish from jumping time after time; then the weight against him begins to tell. Gradually you get him nearer and nearer, until he catches sight of you; then off he goes again, ending his rush with a grand somersault—fish and lead falling back into the water with a splash which you think must shake the hooks from their hold. But the line tightens again, and in a few more minutes you have drawn him up to the side of the boat, where your gillie can use the gaff.

Occasionally when trolling—especially if you are using coarse tackle, or if it is a very calm day—a fish will follow the bait for a considerable distance, pulling gently at it every few yards, but never taking hold. It is very exasperating, and you can do nothing; it is very seldom you will kill a fish of this sort. He has probably been hooked before—at any rate he has seen something to arouse his suspicions, and eventually he will leave the bait. In trolling, a good gillie, who knows the Loch thoroughly, is an immense advantage. He will know exactly how near he can go to certain points without getting the baits fast in the bottom.

In whatever Loch large trout are found, there are always certain points and bays which they frequent in preference to others. These should be tried several times over, either nearer or further from the shore each time. On rough days *ferox* retire to the deepest water, while on moderately calm ones they will be found near to the shore, and sometimes in the bays. Occasionally they are taken in mid-loch, but I think this is only when by chance the baits have passed near some sunken rocks. A small parr about three inches long is the very best bait you can use; for the natural bait, when made to spin properly, will always beat the artificial.

BASS-FISHING.

By G. A. Thring.

From the sportsman's point of view sea-fishing is generally either wearisome or monotonous. It is wearisome to tack up and down all day with a line lazily dragging outside the boat, even though the day be fine and the air refreshing; it is monotonous, when at last the right locality is found, to pull in the line incessantly, with one or two fish attached to it every time. Indefinite slaughter is unpleasant and unworthy of the true sportsman. Bass-fishing has, however, a decided fascination. It is most fascinating, perhaps, when indulged in from the rocks, but it is not unpleasant from a boat on a breezy day.

A coast of mingled rock and sand is perhaps the best ground along which to fish, and the equipment necessary is that of any ordinary salmon fisher, namely, a light salmon rod, a stout line, and either a perfectly white, or a rather gaudy, fly. The weather should be bright and sunny, and the airs brisk and variable.

It is often possible, therefore, for the keen salmon fisher, when the sun is too bright, or the water too low for the king of fish, to take an off day among the Bass. Let him wander over the weed-covered rocks until he finds a spot where the broken strata running out in two long noses into the sea, leave between them a deep sandy ravine up which the tide can run and swirl at its leisure.

Choosing the side which gives him the best chance with the wind, he should walk down to the extreme end of the nose and throw close to the rocks, but over the sand. Along the coast of Cornwall, Devon, and Wales there are many such spots, and the fishing is capital—not too brisk to take away the attraction of mental exercise and skill, and not too slow to make a man give up in despair. Perhaps in five minutes there will be a rise. Do not strike hastily; but when striking, strike with decision. There is a short fight, lasting sometimes three or four minutes—for the Bass is not very game—and then with a landing-net there is no difficulty in lifting out a fish of five or six pounds weight. On a good afternoon half a dozen may be landed in this way, and the fisherman may go home well content with a full creel.

There is another mode of catching this fish which also affords good sport for an off day, and that is trolling from a boat. In this case, again, the weather must be bright and sunny, with a fresh breeze blowing.

The equipment necessary is also very much the same as in the former case, but instead of a fly, a sand eel on a large Steward tackle, a large red indiarubber worm, or even a spoon bait must be the lure. For boat fishing it is needless to say that not only is a good fisherman essential, but also a good sailor. Hire a small sailing boat and a man to attend to it, a man, too, who knows the ins and outs of the coast, and can avoid the sunken rocks. This knowledge is of importance, as the fishing ground is all along the edge of the coast, and a false move, when the boat is well under way, may have disastrous results. Again, the coast must not be

BASS-FISHING

too rocky ; and as a fair example of well-known Bass resorts, the stretch from Westward Ho! towards Clovelly might be quoted, or, again, the shore from Aberystwyth, in Wales, running northward to Borth. Of course, boat-fishing is neither so scientific nor so exciting as fishing with the fly, though with light tackle a good fight is by no means uncommon, and the chances are equalized. Let no man think that because the fisher folk have never heard of such a thing as Bass-fishing along the coast that, therefore, the Bass does not exist in those parts. A keen sportsman can generally discover some kind of amusement even in the dullest of seaside resorts, and the pursuit of the Bass should be by no means the last on his list.

Another pleasing feature is that no costly fishing tickets need be paid for; no tips to keepers need be forthcoming ; there is no danger of fishing in preserved waters ; and there is no expense, at any rate to the 'longshore fisher, beyond that of equipment. Such a sport should be popular, but it has not received the attention which is its due. This is, no doubt, a matter of congratulation to those who, with limited incomes, especially appreciate its attractions. But there is another aspect of Bass-fishing distinctly in its favour. It is a healthy sport, and one without many of the disadvantages of other pursuits. It needs no wading—a frequent cause of rheumatic troubles. It needs no rain and showers, dear to the soul of the trout and salmon fisher, but dangerous to his lungs. The dangers present in the hunting-field are absent here. Bright sunshine, fresh sea air, and plenty of ozone are its chief associations. Truly it is an ideal sport for the worn-out man of business and the jaded city hack.

JULY.

OTTER HUNTING.

By Aubyn Trevor-Battye.

Otter hunting is sometimes decried by hunting men—by men, that is to say, who only hunt to ride—but never, I think, by genuine sportsmen. So far as the science of hunting goes, the pursuit of the Otter demands at least as much knowledge, skill, and experience as the pursuit of the fox. Perhaps, indeed, even more, because of the element in which the Otter moves. It is much as if you were to hunt a fish with hounds; and so I really think there is more craft required, more inference from indications of the smallest, and from general knowledge of the creature's ways.

The Otter is a thing of mystery. It is so greatly nocturnal, it moves so quietly, it shows itself so little, that it may and often does haunt a spot for years without betraying its presence to any but the most practised eye. It will sleep day after day in a faggot stack, in the thatch of a building, under the barn floor, in many another place of this kind, and never be found out. And Otters are great travellers. They will travel miles in a single night, sometimes crossing right over intervening country, hill or valley, it matters not, on their way from stream to stream. If an old dog Otter has been seriously disturbed by hounds, he will generally start away that very night down stream, going on and on, often without pause, till he reaches the big river, or the tide-way, or the sea. For the common Otter is perfectly at home in the sea. The Otters which frequent the caves of our coast, and are commonly spoken of as "sea" Otters, are not sea Otters; we have no sea Otter. They are only our common Otters staying by the sea.

And now a word about the hounds. Everyone is familiar with the appearance of the old rough Otter-hound, if only from the celebrated picture in the shops. Every master of Otter-hounds would, no doubt, like to have some of these beautiful hounds in his pack, but the fact is this: the fox-hound does the work better. The Otter-hound's music, his bell-like voice, is beautiful; his appearance is most picturesque, his nose is wonderfully fine; but, as against all this, you cannot quite trust him. He has a tendency to throw his tongue too freely, to speak without fair warrant. And that is bad. That is why you find, as you commonly do find now, an Otter-hunting pack, if not composed entirely of fox-hounds, at any rate with a preponderating fox-hound element in the pack.

Well, now, I really think the best thing we can do is to go out Otter hunting. Practical experience is the best school. Hampshire is more reachable from town than Devonshire; to Hampshire we will go for convenience, and with Mr. Courtenay Tracy's pack for choice. The meet is at the mill. Time seven o'clock. A perfect morning. Still, clear and chill, the grass drenched in dew. The field is not a large one—it will be larger later on—but early as the hour is, it includes several ladies. If fair faces and bright glances can ensure it, then the master may command success. Hounds work along very quietly for a bit as we move up stream. But presently a hound opens, and another follows suit. Is it all right? These are not exactly the hounds one would trust. There, that is better, that is old Dreamer's voice. Dreamer does not make mistakes. The pack knows this well enough, and in an instant Dreamer is the centre of a dozen waving sterns. Every nose straining to pick up something of that perfume which set Dreamer's tongue a-going. But scent is a kittle thing; one hound will own to scent of two days old, another will not speak where an Otter must have passed not many hours before. However, there is no doubt about it this time, for just above the shallow at the tail of the next pool, hounds suddenly break into a pretty chorus. Is he up or down? The master does a wise thing. He is in no hurry. He deliberately takes hounds off down stream and makes a cast or so where the two big "carriers" come in just below. But nothing results; he can, therefore, feel that he has made all good. Hounds can get to work again—they are not running heel. It would only be wearisome to describe the hunt in detail. Suffice it to say it is one of the prettiest drag hunts ever seen. Scent gets hotter and hotter, and after every momentary lull it is beautiful to see the way hounds flash again to the line. See there on a spit of sand is his "seal" or footprint—quite different from a dog's—five toes instead of four. After an hour or so at this, hounds come to a halt. Yes, they are baying their Otter, he is somewhere up under the roots of this old tree. Put in the terrier? No, we will try to move him first by other means. This is soon accomplished. "Gentlemen will kindly jump. Now then—all together!" And about a dozen men having closed up above the hover, throw all their energies into this performance. Such a shaking is too much for our Otter. No one sees him go, he slips out so adroitly; but gone he is, as the hounds can tell you. Tally ho! as he slips down the stream, he is viewed by a watcher stationed by the shallow below. And then begins

OTTER HUNTING.

a proper hunt. Hounds are mad. They rattle him, they bustle him, they give him no pause. But he presently gains a long deep pool, and here for a bit he has it all his own way. And very pretty indeed it is to see the swimming hounds as they take the scent off the top of the water. But he is forced to quit at last, and crosses the open, viewed by everybody, right across the point of a meadow and into a second stream. But here his fate is sealed. There is not much holding here, and in half an hour's time hounds are all on the top of him in about a foot and a half of water. Above the pool and below it the shallow is bound by a human line standing shoulder to shoulder, foot to foot, so that scarce a water-rat could find room to pass. Even so it is absolutely marvellous to see how long the Otter can evade his foes. No one tailed him. The hounds got him fairly enough. There he is out on the grass. Who-oop! "Ladies and gentlemen, three hours and a half, and twenty pounds if he is an ounce." Now for the trophies!

SEA FISHING FROM PIERS.

By E. T. Sachs.

If no other method for the division of mankind were available, that of anglers and non-anglers might be adopted as a rough-and-ready one. Certain it is that the angler is born and not made, and also that one portion of the human race comes into the world so constituted as to entertain a loathing for the pastime of angling, in any form whatsoever, until the day of its death. This is the portion which, on the occasion of its visits to the seaside, exhausts its vocabulary in seeking for terms of commiseration with the other portion that finds the sum of human happiness in fishing from the piers. It is not every angler who will pursue his favourite pastime so far as this; and, indeed, it must be conceded that one must have the angling mania very pronounced to undertake to angle from a crowded pier at a fashionable resort. By a fortuitous dispensation, however, it happens that the best fishing is not obtained from the most fashionable piers, but from those belonging to the more retired seaside places. At both Margate and Ramsgate, which are scarcely noted for the retiring nature of their summer and autumn visitors, pier fishing has been occasionally practised with success, it is true, but these are not typical spots for the practice of angling. As a matter of fact, one would hesitate to indicate any particular spot where the pier fishing is particularly good, because the sport often varies with the season ; but I know of no place within reasonable reach of London where one can better rely upon a bag of some kind, virtually all the year round, than at Deal, situated between Ramsgate and Dover, in Kent. The tide at this place sets past the head of the pier, going either north or south ; and, whilst one state of the tide is better than another, there seems to be no period of it when the fish absolutely decline to feed at all, though the thick water, full of sand in suspension, which prevails after stormy weather, will of course put fish off.

I have mentioned Deal because I have often fished there, and anglers may now reasonably live in hope that Dover will be restored to them as the first-class angling resort which it once unquestionably was. Its decline in this direction was contemporaneous with the building of the Admiralty pier; and the hope for the future lies in the recent completion of the new pier, which, extending three hundred yards out, may be expected to take the anglers amongst the fish. There are, of course, many other piers round the coast from which fishing may be had, and the methods adapted to one are adapted to all.

I may say at once that the hand-line fishing which in most cases, though not in all, is the best method when fishing from a boat, is not the proper one to adopt when angling from a pier. The amount of sea room at the disposal of the angler from the boat renders the absence of control he has over the motions of the fish, beyond hauling him neck and crop on board, a matter of indifference. When his line is baited all he has to do is to drop it overboard; for the pier angler it is at least necessary that the bait should be kept away from the piles, and this cannot be managed without the aid of a rod. As a rule, it is advisable that the bait should be some distance out, and here the rod becomes invaluable, for by its aid the bait can be cast with greater ease and accuracy than by the hand.

The angler who comes fresh from his fresh-water fishing will naturally wish to know to what extent his tackle and practices have to be deviated from. As regards tackle, the conditions in the sea are very different from those obtaining in fresh water, where great lifting power is not demanded as it is in sea fishing, owing to the many times increased weight of the paraphernalia. If the angler is possessed of a very stiff bamboo pike rod he may bring it into service; but he will find it best to have an article made about nine feet in length, and very much stronger and stiffer than is ever made for pike. A Nottingham reel, with a check, should be used, and as the line must be stout, a large reel will be necessary. The rings of the rod should be large and of the "snake" pattern, the top ring being furnished with a little pulley wheel. This special top ring for sea fishing is supplied at the tackle shops. I am not saying that a pike fisherman could not take his tackle direct to the sea and use it successfully, because this is frequently done. But anyone intending to take up sea fishing will find it advantageous to have a separate outfit, the work being too trying for tackle made for another purpose. A strong pike line will hold any fish that is likely to be hooked from a pier, though cod and lythe of ten pounds and over are not uncommonly caught, and the rod will prove equal to playing them.

SEA FISHING FROM PIERS.

Two methods which are popularly adopted are the "chop-stick" and the paternoster. The "chop-stick" is the cross-piece of whalebone (about fifteen inches is a good length), suspended from the line by the centre, and having depending from each of its two extremities a length of gut with a hook at the end. Other materials beyond those mentioned are employed, but these give the best results. People who know no better, laugh at fishing fine in the sea, but experiments have proved that under certain conditions, which are not controllable by the angler, fine tackle beats the coarse. Depending from the centre of the "chop-stick" is a line with a heavy lead attached, which lead is held a foot or eighteen inches from the ground. As the lead will weigh a quarter of a pound or so, we see at once the necessity for a stout rod. Some anglers adopt a four-ended "chop-stick," made by binding one piece of whalebone at right-angles across another. In this way four hooks are used at once, and no doubt greater chance is given of a bite. Nothing is more mysterious than the way a hook is denuded of its bait when sea fishing, without the slightest indication of any molestation of it being afforded, and it is very annoying when hauling the line in, after a protracted period of quietude, to find both baits gone. Contemplation of the length of time during which the angler may have been waiting with a baitless hook is never a satisfactory occupation. The angler with four hooks instead of two is less liable to this irritating episode ; but he requires a stouter rod.

The paternoster fisherman requires a heavier lead in the sea than he does in fresh water, the depth being so much more considerable, and, consequently, the weight of water against the line so many times increased. He may have as many hooks as he pleases, or as there is room for, but he will find three sufficiently troublesome to manage. The lowest of all should be very near the bottom, for the benefit of the flat-fish, and the others only so far apart as to clear one another. For paternostering at sea the tackle shops supply an arrangement consisting of a short length of brass wire, swivelled, and with a loop at either end. From the centre extends an arm of stout brass wire, with a ring at the end to which the gut hook length is attached. The contrivance is inserted in the line where required, and it has the effect of causing the baited hook to swing out with the tide, whilst it also prevents any fouling or kinking of the line when a fish is hooked, no matter how many times it may circle round.

The lead of the paternoster lies upon the bottom, and the least attempt to molest any of the baits is felt by the angler. The lead is thrown some distance out, and in this way far more ground is covered than can be managed with the "chop-stick" arrangement, which is confined to the immediate vicinity of the piles of the pier. On a bite being felt, the angler should strike, and strike hard, whatever be the tackle that he is using.

In conjunction with the paternoster, a leger bait may be used— that is to say, an extra hook below the lead which lies upon the ground. In this case a pipe-shaped or flat lead, with a hole through it from end to end, must be employed. The line passes through the hole, a knot preventing it coming too far, and when a fish seizes the bait lying upon the ground the line is drawn through the lead and the fact communicated to the angler.

If the water is still, float tackle, such as is used in rivers for perch, may be employed, and whiting will be caught. Grey mullet, which are a very shy fish, frequenting still places such as harbours, require fishing for with the greatest circumspection and with a special tackle. This comprises a long length of medium-sized gut, having three or four small pieces of cork attached to it (a slit along the cork enables this to be done easily) at intervals of a couple of feet. Depending from the gut are three or four hooks, which are baited. This arrangement is allowed to drift away with the tide, surrounded by stale bread crumbs and small pieces of crust wherewith to attract the mullet.

Fishing with a white fly sometimes answers from piers; and at Deal at a certain period of the year, the autumn, pieces of parchment in the shape of fish are attached to the hook, and danced on the surface. These are taken by the coal fish.

What baits are used will depend very much upon the locality. The lugworm is a universal and safe bait, though not easy to put upon the hook, and not always obtainable. The hunt after it is a sport in itself, so quickly does it disappear into the sand. Where there is much fishing, the lugworm when procurable at all can generally be purchased from boys. The mussel is also a universal bait, no doubt because so readily obtainable. Oysters are also a good bait, but, of course, the price of any but the commonest makes them prohibitive. Pieces of fresh herring or mackerel, shrimps, prawns, limpets and whelks all do good service. On certain coasts the sand eel can be obtained. This is the best bait of all, but it is difficult to get, and must be kept alive in a creel. When obtainable, the cuttle-fish, or squid, is a good bait.

AUGUST.

THE WHITE TROUT.

BY G. H. THIRING.

THE White Trout is, without doubt, to use the expressive language of the prize-ring, the gamest fighter for its weight of all the migratory fish in England, or the stationary inhabitants of our lakes and rivers either. Its fighting weight runs from one and a half to five or six pounds. The way to test his fighting powers is as follows :—The fly-fisher must tempt him with a carefully chosen and not too gaudily coloured fly, and must make his cast with equal care in a likely run. If the fish be at home and hungry, a rise will most probably follow, and then there is scarcely need for the usual strike before the hook is securely fixed and the game begins. With a rush and a leap the Trout is out of the water, and the wielder of the rod, especially if he is fishing with a fairly light rod and tackle, will have to do all he knows to keep the Trout at the further end of his line. He drops his rod point as the silver body splashes into the water, and he must be ready to feel him gingerly and guide him carefully in his maddest rushes. It is not so easy. Suddenly there comes a tug and a bolt. This is the point at which to be ready. The proverbial fool who, according to Dr. Johnson, presides at the shore end of the rod and line, deserves his name if at this point in the game he break his tackle. If he does, his line rebounds with a swish about his ears. He swears, unless he be philosophically inclined. If he is that, he sits down, replaces what he has lost, and starts again a wiser man. The old hand is prepared for the emergency I have described, and pays out his line gently and carefully till the strain is slackened and his game opponent is under control.

The sport, however, is by no means at an end. Even when the game seems won, there may be two or three such leaps and rushes, or, what is almost as bad, a clean bolt towards the shore, when the greatest rapidity is necessary to draw in the line, and keep a steady strain on the victim. If fortune is favourable, after as good a quarter of an hour as any true-hearted sportsman could wish for, the fish may be brought gently and with delicacy to within reach of the fatal net or gaff.

There are many places in England, Scotland and Ireland where White Trout fishing is excellent. The Shetland Islands, the Hebrides, many of the Scotch and Welsh rivers, and many of the lochs and rivers on the west and north-west coast of Ireland, can give a good account of themselves in this line of sport.

To obtain good sport, it is best to fish with nothing heavier than a strong trout rod, and trout tackle. The fish will, indeed, come readily to spinning minnows and small trolling bait, but this method should be the last resort of a true sportsman. There are those who try to tempt the White Trout with rod and line worthy of the more lordly salmon, and no doubt with some success, but they lose the excitement and enthusiasm of the true artist, the hopes and fears caused by handling a heavy fish on rather light tackle, and lastly, the chance of filling their creel on a day unsuited to the gentle pursuit.

On this latter point, the writer speaks from personal experience. He had been fishing in one of the Welsh rivers for salmon and White Trout, and each successive day the odds had grown in favour of the fish, and against the fisherman. The sun was provokingly bright, and the river had become as provokingly low. The fishing party had started out as usual, consisting of two or three veterans, with large salmon rods and tackle, and with gillies, whose faith in the powers of their masters was only equalled by their ignorance of the true art of fishing, and the writer unaccompanied, with a moderately strong trout rod, and good-sized trout flies. His departure was not imposing, and he was diffident as to his own powers. But his return, with full creel, when the others had not even had a rise, was triumphant and satisfactory.

Among the fishing resorts mentioned above, the Hebrides can strongly be recommended; Loch Boirdale in South Uist is a splendid camping ground. But the accommodation, though good, is limited, and steps must be taken to secure rooms beforehand. The low-lying plains of the islands are intersected with lochs into which the White Trout run, fresh with the vigour of the sea. The best fishing grounds lie some way from the hotel, but conveyances are ready to take the fisher every morning. As there is no other sport but fishing on the islands, the companionship is usually congenial, and there is no difficulty about arranging for the different fishing grounds. The cost of living, etc., is cheap and the fare is good, and, beyond the initial cost of the journey, the expense is not great.

WHITE TROUT FISHING.

These remarks apply in a general sense to the Shetland Isles and to the fishing in the north-west of Ireland.

There is one drawback to the fishing in the Hebrides, and that is the presence of flies—nothing worse than the common house fly, but in swarms. It is constantly present with the aggravating hum of its myriad wings. It irritates the calmest of men as he sets up his rod and adjusts his tackle. Flies spoil his pet cast and blind him as he is landing his largest fish. Otherwise they are harmless. There is one loch on South Uist that is called by an unpronounceable Gaelic name, which means the loch of flies. This loch is full of trout, but the followers of Walton are warned against trying it.

This drawback, however, is but a small one. Were the harmless fly changed to the more terrible mosquito I doubt whether the true sportsman would be held back from these waters.

CHUB FISHING.

BY E. T. SACHS.

IT is a fortunate thing indeed for the Chub that it is the least toothsome of all fresh-water fishes. Were it as much an edible dainty as it is the reverse, its extermination would be a mere question of time, for certainly a more omnivorous fish does not swim, nor one which adapts itself with greater complacency to every kind of water. Of most fish it can be said that they have their particular fancies and are to be found only in certain portions of rivers according to the flow of the water, but the Chub is everywhere. No depth or condition of water comes amiss to it, and the angler is equally likely to catch one when spinning for trout in the tumbling waters of a weir, in the depths of a barbel swim, or in the seclusion of overhanging boughs. I have caught numbers with the fly in the stiller portions of the crystal trout streams of Luxemburg, or in the Continental rivers, with the cockchafer and grasshopper, and I have seen the native angler, using worm or pellet of bread paste, drag the fish out from near sewage outlets, much as small roach used to be caught by the hundred at the Kingston sewer. To go still further afield, I have caught the Chub (nearly the same species as our own) in the

Highlands of Sumatra, where the Malays preserve the fish in private ponds. Peculiarities in the methods of feeding these fish rendered one averse to eating them. The Chub will feed on anything and everything, from gentle to live bait. Nevertheless, there are certain methods which experience has proved to be more efficacious than others for making bags of this fish. Although to be found in literally every part of the river, he haunts certain places in larger numbers than elsewhere. Fish follow their food, and when the hungry period which immediately follows the spawning time, and which drives the Chub to the weirs where small fry are congregated, is over, quantities of the fish settle down in "runs," where the flow of water carries down food in a more or less narrow channel. Others, again, distribute themselves where overhanging trees promise contributions to the larder in the shape of caterpillars, beetles and other insects. On the approach of any disturbing element, such as a boat on the water, or a pedestrian on the bank, the Chub, who is foraging under trees, immediately bolts to a refuge under the bank, which is generally more or less hollowed out ; or a convenient root will serve the purpose just as well. Poachers know this propensity of the Chub, and wading in the stream, feel for the fish under the bank.

The shyness of the Chub is extraordinary. A falling leaf will send a shoal flying in all directions, whilst the smack of an oar upon the water, will drive them perfectly frantic, and nothing will be seen of them again for hours. Therefore, whatever be the method you may employ, one general rule that must be observed is the adoption of extreme caution. Except high up, above Oxford, where the river is narrow, it will be necessary to fish from some kind of vessel in order to reach the fish under the boughs by means of a fly rod. If the angler be skilful with the Canadian canoe no craft is more suitable, so very little disturbance of the water being occasioned, whilst the fisherman is very low down on the water. Once within range, all the paddling that is necessary can be done with the left hand, and when a fish is hooked it does not much matter for the moment what becomes of the canoe. The next best craft is a sculling boat, one occupant sculling and the other fishing. The sculler must learn to do his work noiselessly, and when the boat is upon the scene of action, the sculls should not be taken from the water at all. The angler should face the stern and be gently backed towards the place where the fish may be supposed to lie. A punt may, of course, be used for the work, but it will be better if a paddle be employed instead of a punt pole.

Although the angler is using a fly rod, it does not necessarily follow that he is fishing with the fly. In the quiet portions of the river very good sport indeed may be had with the fly, a red or black palmer being as good as anything, especially if a little piece of wash leather be added as a tag. But it is not always that the Chub are near the surface, and then a bunch of gentles or a small frog, which will sink deep into the water, is more efficacious. The bunch of gentles should not err on the side of economy, for nature has provided the Chub with a large mouth. The frog is used dead, and is strung upon a hook, head downwards, the hind legs being bound to the hook shank. To this bait some motion should be communicated when it is in the water; and the Chub, fortunately, do not take cognizance of the unnatural conduct of the frog in progressing backwards. All they see is a frog, and are

CHUB FISHING.

bent upon having it. A practised fly fisherman will adapt himself to these methods in a very short time. To the tyro one can only say that it does not matter how heavily the fly, bunch of gentles, or frog alights on the water. The chief thing to be attended to is the keeping oneself out of sight. As Chub lie very close under low hanging boughs, some little risk must be run in endeavouring to throw under such. When a Chub is hooked it must be held, or it will probably rush behind a root and be lost. The first rush of a big Chub is very powerful, so the tackle must be strong; but after the first rush the fish generally gives in, though in lively streams I have seen a Chub afford excellent sport.

A cockchafer, or two or three grasshoppers make capital baits for Chub, but throwing them à la fly is a delicate operation. As they do not sink they may be allowed to float when once thrown, and if there be any stream it may carry them over fish.

Fishing for Chub with the fly rod is certainly more adapted to the upper than to the lower portions of the Thames. For fishing these last-named, no style of angling is more adapted than the Nottingham, which is conducted with a free running reel, carrying perhaps a hundred yards of fine line of twisted silk, with a large quill or cork float. To use this tackle with success it is by no means indispensable to fish from a boat, if the angler can meet with places where the bank juts out and forms a narrow stream, or where two streams converge; but the employment of a craft of some kind confers great advantages in enabling the angler to shift his ground, which is very necessary in Chub fishing. No experience is more common in angling for this fish than to catch one of a shoal at once, and later on to discover that the remainder have taken fright and departed.

As is the case with every fish known to me, a variety of baits are recommended for Chub, but I am quite prepared to pin my faith to plain and simple cheese. It is as well to select a "strong" cheese, the flavour of which is disseminated in the water and whets the Chub's appetite; but old cheese has the serious drawback of soon becoming very hard, and so interfering with the strike. A piece is worked in the fingers into a paste and compressed round the hook. With the punt fixed at the head of the "run," the bait is allowed to follow its course. As a rule, the stream will be powerful enough to carry the line off the reel, but the hand must be ever in position ready to touch the reel when the even progress of the float is in danger of being arrested, for it is essential that no jerks be communicated to the bait. The theory upon which the admirable Nottingham style of fishing chiefly depends for success is embodied in the motto, ",fine and far off," and by its means the Chub fisherman fishes with a light heart, thirty, forty, and even fifty yards away. Great care must be observed, for at these distances the line may easily become slack, in which case it will not be possible to strike the fish in the event of a bite. Therefore, have the forefinger on the reel, always keeping the line taut. The light silk line does not sink and so is always in sight. The farther the bait is away from the angler the more pronounced must be his strike, and the instant he has struck he must begin to wind at the reel, so as to immediately hold his fish. Then he can pause for an instant to see what his captive means to do. The elasticity in thirty or forty yards of line is very considerable, so, with a quantity of line out, the danger of a breakage from the first rush of the fish is not great, especially as it will not necessarily dash headlong down stream, but more probably make to one side or the other.

A very essential adjunct is the ground baiting. I call it ground baiting for convenience, but strictly it does not comply with all the conditions necessary to qualify for the title, since it is not intended to reach the ground. Just previous to the bait being allowed to swim down the "run," a piece of cheese is taken in the hand and pared off into the water. Cheese parings are by far the best form in which to bait with this article, as they sink very slowly and make a good show in the water. What the angler hopes is that the Chub will be attracted by the shavings, and feed greedily upon them, and when the angler's fat lump upon the hook is seen to approach, a rush of many fish will be made for it, the successful one being the largest of the shoal. Three or four lots of cheese shavings, and the same number of swims, may be considered to sufficiently test any run; there is no advantage in remaining an hour or two at the same spot, as is the case with roach. There is not much more to be said, beyond the advice to fish as fine as you dare, and to place the shot necessary to cock the float, and keep the bait down at least eighteen inches from the last named. With a light and pliant rod, such as is manufactured for the purpose, finer fishing can be indulged in than with a heavier rod, heavy rods being, as a general thing, a mistake. The rod I use for Nottingham fishing I have, on occasion, used as a fly rod to cast a worm for rudd. I mention this to give a notion of its pliancy. The great thing is to have the pliancy distributed throughout the whole length of a rod, and not merely through the smaller joints. Such a rod plays a fish of itself.

Sometimes shoals of Chub may be seen in exposed positions. I know of no better method for catching these than a minnow on roach tackle, the minnow a couple of feet only from the surface. If the angler captures one of a shoal at one attempt he must be satisfied. If he effects such capture and then looks for the remainder of the shoal, he will, in the majority of cases, look in vain.

CHAR FISHING.

By H. A. Bryden.

Of all our fishes the Char is, perhaps, the most fascinating, whether from the æsthetic or from the sporting point of view. The wonderful colouring, sometimes crimson pink, sometimes ruddy orange, with which the lower parts of its silvery body is suffused, the rarity of its occurrence and the extreme difficulty attending its capture, render this fish not only an object of unique interest to the naturalist, but also of pleasure to the angler, in the rare and fleeting moments when it is in the vein and will come briskly to the fly.

In these islands the waters in which Char may be found are to be counted almost upon the fingers of one's hands. In the lake district, Windermere, Buttermere, Lake Ennerdale, and Crummock Water, all contain fair numbers. In Ulleswater, which formerly held Char in abundance, the fish is quite extinct; and of Coniston, thanks to its pollution by the copper mines, the same miserable tale is to be told. A few waters of Wales, chiefly in Merioneth, still support Char, while in Scotland in the depths of Loch Tay, Loch Fewin, Loch Earn, Loch Roy, Loch Lubnaig, Loch Inch and a few other lakes, this most brilliant member of the *salmonidæ* still abounds. In Ireland the Char is to be found in Lough Esk, Lough Melvin, and in various deep waters in Cork, Donegal, Wicklow, Galway, and Fermanagh, and in the room of the Piscatorial Society, Holborn Restaurant, are to be seen stuffed specimens from another Irish lough.

The true home of the Char, however, is in Northern Europe, and especially in Swedish Lapland, where immense numbers of these fish may be taken with fly or minnow, if the angler is prepared to venture thus far and to brave the intolerable swarms of mosquitoes and gnats that, from June to September, wreak their bloodthirsty wills upon the invader. Char grow here to a large size, as they do in the waters of Iceland, five pounds being a not uncommon weight.

For a long period there was much confusion concerning this fish and its habits. Even now it is known indifferently as the Case Char, Red Char and Gilt Char. It is well-established, however, by this time that in its normal condition, when showing least colour, the fish is called, locally, the *Case* Char. Before spawning, when it assumes its most vivid crimson, it is the *Red* Char; while at another season, having shed its spawn, its colouring changes to a metallic orange, and it is known as the *Gilt* Char. Dr. Günther, after extensive research, has distributed the British Char in five species; but these are little recognized by the outer public, and indeed, even among some scientists, are thought to be merely varieties of one species.

For some centuries potted Char has been a famous delicacy in the Lake country. Defoe seems to have known of this dainty, and says, speaking of the fish in " Winandermere" (Windermere), "it must needs be a great rarity since the quantity they take even here is but small." But long before Defoe's time, "char pyes" and "potted charrs" seem to have been known to Elizabethan and even earlier *gourmets*.

The Char, like its cousin the yellow trout, is non-migratory— that is, to the sea; it loves rocky bottoms, and, as a rule, the deepest and coldest waters. It is said that the fish retire for spawning purposes to shallower parts of the lake, and even to rocky streams, in November, returning to their well loved deeps towards April. It was during the spawning season in these shallows that such numbers used to be captured with nets by the Lake people concerned in the "potting" business. When safe in deep water their capture is far less easy, and the angler, "charm he never so wisely," will find a brace or two a day an average basket.

Spinning very deep with the minnow is perhaps the best course to adopt. The "phantom" will occasionally, in the Highland lochs, account for an odd fish here and there. Towards May the angler may expect to have some little sport by these methods; and between now and September he must make up his mind to catch his Char with fly or minnow, for he will get them at no other season. A more deadly plan is to trail a long and heavily weighted line, baited with minnow, behind the boat, which is then rowed very quietly about. Another lure occasionally used— almost as poaching a contrivance as the "otter"—is to let down a weighted line, upon the lowest part of which are fastened at intervals artificial flies. But, after all, these are only the baser resources of the pot-hunter, or rather of the "potter" who must have his Char by fair means or foul.

In Britain it must be sadly acknowledged that it is only by the rarest chance the fisherman may hope to catch the Char in the mood for the fly, and then usually only in the months of July and August, when he comes up towards the surface for the pleasant warmth. The minnow, of which this fish is, like many another finny *viveur*, inordinately fond, is best calculated to tempt him in the depths. I have never tried Char with the paternoster, but there seems to be no reason why that killing method in combination with a tempting minnow or two should not be successful. The great depth would no doubt tell somewhat upon the delicacy of touch so requisite in this style of fishing. So little indeed are Char reckoned as sporting fish, and so desperate seems their pursuit, even to the most skilled anglers, that Mr. Cholmondeley Pennell, in his volume of the Badminton Library, dismisses the fish with scant notice. "They are," he says, "unfortunately so seldom captured by the rod and line, that they are objects more of interest to the icthyologist than the fly fisher." And yet on the rare occasions in one's angling career when these most beautiful fish are in the humour, they will take the fly as greedily as the hungriest trout. Sometimes on a warm day in May or June, more often in July and August, will this happen. Such red-letter hours form scarce memories indeed, and will be recalled with regretful pleasure for years after. I shall never forget an afternoon on Loch Tay a few years back, when, quite suddenly, the Char emerged from their seclusion, and for one fascinating spell of forty minutes came repeatedly at my trout flies.

We had pulled out from Kenmore past that lovely wooded islet whereon two queens of ancient Scotland lie in their quiet graves. The time was summer, the day soft and gleamy, the atmosphere perfection. All around us sprang the very fairest of Highland

CHAR FISHING.

scenery. The dark forest of the Drummond Hill, the swelling uplands above Acharn; the noble expanse of lake—seven or eight miles of it in sight; on the right hand the towering mass of Ben Lawers; wherever the eye roved a combination of mountain, wood, and water hardly to be excelled in any part of the world. At first we used the blue phantom minnow, and, pulling slowly along the southern shroe, were fairly successful among the trout. After lunch a good breeze sprang up, there was a strong ripple upon the water, and the flies were got out.

Presently we crossed to the northern shore, between Lawers and Kenmore, and I began to cast my fly just where the deep water meets the shallow. The trout were, for a time, ravenous, and for something less than an hour the sport could not have been bettered. For the Char, too, the charm was broken. They came at the flies freely and added a wonderful zest to the sport. As a rule I found that they took the fly when well under water; occasionally, however, one caught the flash of a ruddy belly as they rose to the surface; and what a keen pleasure it was as one drew the living bars of clean, pinkish silver, after a sharp tussle—for they are strong, active fish—into the landing net. They were not particular about the pattern of one's flies, though in point of fact red palmer and coch-y-bonddhu did most of the work. For half-an-hour or so they came, and then their noble ardour suddenly abated. Among the goodly basket of trout, one of the best on Loch Tay that season, I had eleven Char averaging nearly half-a-pound apiece. The biggest fish was a shade under a pound in weight. This is not an extravagant catch; yet for this shy fish it may be reckoned good, in these islands. The Char, it may be added, seldom, if ever, exceeds (in Britain) two pounds in weight, and, more often than not, runs some three to the pound. But whether the angler be pulling leisurely about fair Windermere in the hope of enticing this beautiful fish from the deeps, or trouting, with the off chance of a Char, beneath the shadow of Ben Lawers, or upon some wilder and more remote northern water, he can never lack for pleasant days and matchless scenery: and, perchance, some fine breezy summer's day he shall find this goodly fish eager and in the humour, and his measure of happiness will be temporarily complete.

THE HABITS OF THE WILD RED DEER.

By H. H. S. Pearse.

WHEN the wide tracts of ridge and combe bordering Exmoor Forest are brightest with the glorious hues of purple heather and golden gorse; when every frond of brake-fern is unfolded, rising in hardened stalk high enough to shelter all but the branching antlers of a lordly stag, the natives of that sporting county care to talk of nothing but wild Red Deer and their ways. Between bracken and deer there is closer association than any but keen observers of nature know. In the early days of winter, when the ferns are brown and easily broken because the sap is all out of them, stags retire to woodland recesses and shed their antlers among the fallen leaves. When tiny green volutes begin to shoot up again in spring-time, new antlers are sprouting on the heads of male deer. So long as bracken bears on fronds and stalks a downy coat, the stag's antlers are in "velvet," and sensitive to every touch. A russet covering, like dry moss on an apple tree, encases the branching growth of nerves and blood vessels, to protect them from the flies that swarm in deer coverts, and from being hurt by trailing brambles. This "velvet" is toughest when young, but as summer days lengthen and ferns begin to open into glossy fronds, it may be readily torn if caught in a thorn branch or stout oak twig. From such mishaps the stag suffers much pain, and occasionally the perfect growth of brow, bay and tray is marred thereby. All this time stags, as if conscious that they have neither weapons for defence nor full majesty wherewith to impress the opposite sex with a sense of their power, keep away from the company of hinds, hiding in thickets by day and feeding alone at night. When the bracken has lost all traces of down, and its stalks, hard and polished, are turning a tawny tint, the stag's antlers, hardened into bone, begin to burst their "velvet." It may be seen then hanging in mossy shreds from every tine. To get rid of this disfigurement the stag uses what is known as a "fraying stock." It may be the trunk of a gnarled oak, from which a stag will strip the tough rind as he rubs his antlers against it and tramples the ground round and round; or it may be a sapling, which he will twist into all manner of strange shapes in his endeavour to free himself of the velvet and point the tines for combat. Now, as July draws to a close, is the season for fraying, and one may at times hear the antlered monarchs at work in the deepest recesses of a leafy valley with sounds such as single-stick players make when their feet beat time to the tapping of tough ash wands, and the swift play has made them scant of breath. At night the deer wander out to feed in cornfields or among turnips, or to strip apples from orchards. This is the harbourer's opportunity for learning all about their habits, which may serve him in slotting them to their lairs when the hunting days come round. Without a skilled harbourer, who can tell almost to a yard where the heavy harts are lying when wanted, stag-hunting, as pursued on Exmoor, would be almost impossible. He must lie out at night to discover where the lord of each herd has his favourite feeding-ground, and by what path or "rack" this fastidious gentleman comes back to covert at daybreak. Some fat deer are too lazy to wander far from their favourite haunts; while others have been known to make their way twenty miles across the moor in a few hours for change of diet. It is the harbourer's business to find all this out, though he may rarely want to-draw on such resources of knowledge. The country over which he exercises his craft is as wide as the county of Rutland, and he must sometimes cross it between sunset and dawn to begin his work of slotting, before the faint imprints of hoofs on dewy grass or dusty paths have time to disappear. He will go first into a cornfield, or orchard, or turnip plot, and a glance is enough to tell him whether stags or hinds have been feeding there. A hind draws the ears of corn through her teeth, stripping out all the grains, but a stag nibbles only half the ear daintily. In a turnip-field the hind eats clean down to the ground, while a stag pulls the root up, and after taking one bite throws all that is left over his head in lordly wastefulness. Among apple-trees hinds only strip off the fruit that they can reach easily, but if branches are broken at a height of seven or eight feet the harbourer knows without looking at the ground that a stag has been there, but why, he will probably not be able to tell you. Richard Jefferies in all his researches did not discover by what means stags get at their favourite apples, higher than a tall man can stretch his arm. If you watch an orchard, night after night, however, taking care that deer cannot get scent of your presence, you may, perchance, see a great stag raise himself on his hind legs and then, leaping up, strike at the boughs with his strong antlers, using them as a woodman does a billhook. Thus he gets enough for an ample meal, which he eats so voraciously that he takes no trouble to crunch the apples, but swallows them whole. At the first gleam of dawn red deer make their way to the densest thickets, and to reach these they use, time after time, the same tracks, with which they become so familiar that anything unusual there—a twig cut and laid down, or a string of wild birds' feathers stretched from bush to bush, is enough to give the alarm. If hard pressed in chase, however, a stag does not look for any path, but crashes through the thickest copses and tosses the netted oak branches aside as if they were spray. To do this he lays his antlers back until they rest each side of his flanks, acting as the cutwater of a ship that cleaves through waves and flings foam from her bows. When the harbourer has learned by certain signs that a heavy stag has been feeding anywhere, he proceeds to slot him, and in this he shows skill not inferior to the Red Indian's woodcraft. He can tell by the size of a slot, and the width that the cleft hoofs were apart when they made that imprint, whether the stag is heavy or light, and whether he went at a walk towards the covert or was alarmed into quicker flight. If there has been rain enough to make the ground soft this is easy, but often a harbourer can find no trace plainer than a deer's hoof makes in crushing down blades of grass, so that they look dark green amid the grey dew. At one point perhaps he will draw your attention, if you have the privilege of being his companion on such an expedition, to some mark on hard ground, scarcely perceptible in the dim light of dawn. "That's ov'n shore 'nuff," he will say, "but 'er didn' maake thickee lāst night, I'll warn." Asked how he can read that in a print that looks quite sharp and fresh to untutored eyes when they can quite make it out, he replies, "Why, doant 'ee zee there's a spit of rain in the middle ov'n, and us han't had no rain since yesterday morning." Presently, however, the harbourer finds a slot that is unmistakably fresh leading into covert. Still his work is only half done. He must go all round the covert if it is small, or down its rough paths if it be large—as Exmoor woodlands mostly are—and find out whether the stag he wants has crossed or gone on to another woodland, or whether it is lying near at hand, and all the while care must be taken not to alarm the stag, who if he "winds" the presence of foes will be up and away. By a fern bent down and bruised, a broken twig, a curved line on the hard path scarcely visible to any but himself, or splashes of water on boulders in the little brawling rivulet, the harbourer knows if his quarry has gone further. All this may mean hours of patient work. When once satisfied, however, that he has the right stag safe, the harbourer makes his way out of covert silently and swiftly so that the cunning animal may not take flight too soon. Then he goes off to meet the hounds and guide the huntsman to where the runnable stag has been safely harboured. When roused from his lair, the Red Deer brings many cunning tricks into play before hounds can drive him out of covert, but the shifts and subterfuges by which he strives to elude pursuit or baffle his pursuers must be subject matter for another article.

FLAPPER SHOOTING.

By J. R. Roberts.

"When Autumn, crowned with the sickle and the wheaten sheaf, comes jovial on, nodding o'er the yellow plain," tis pleasant to wander, gun in hand and dog at heel, down some moorland valley, or across the marshy hollows of some upland plateau, in quest of wild fowl. Not so long ago, in July, the sportsman would hunt through the rushes in the deepest and most retired parts of a peaceful brook or prattling trout stream when seeking Flappers. At this season, when the old duck is sprung, the brood is not likely to be far off, and when once found they are then easily killed, as they attain their full growth before the wings are fledged. The beneficent Wild Birds' Act, however, changed all that. Indeed, August is quite early enough for Flapper-shooting; and in September it is often a welcome relaxant concomitant with the delights of the partridge campaign. "When autumn's yellow lustre gilds the world," the young wild duck are still truly, if not technically, Flappers. "When the Flappers take wing," says Colonel Hawker, "they assume the name of wild ducks." In this age the pedantries of the phraseology of sport are considerably modified; moreover, despite the dictum of the greatest authority on the subject, one is justified, even in September, in designating the comparatively recently fledged young ducks as " Flappers," for they do not in their first season achieve the full glory of adultness. As soon as ever the inner side of the wing is fully clothed, they take to flying; their bones, previously of a gristly character, quickly harden, giving the bird full power and use of its pinions. But it is only the full plumaged or older males that exhibit the feathers so useful in fly-making, and which may be seen in nearly all salmon flies. So it is of Flappers and Flapper shooting that we discourse. About August they repair to cornfields, till disturbed by harvest people. Then they frequent rivers, streams, and the wet parts of commons, wastes, dingles and moors. One conjures up a picture of a sportsman, duly equipped, with solitary attendant and steady well-trained dog (for choice a brown Irish spaniel, known to some lips as a retriever), plodding up a lonely valley, hemmed in with rolling moors. Alternately pool and cascade, the rivulet comes towards and past him. The scarlet berries of the mountain ash are finely foiled by the bronzed expanse of bracken waving on the hill-side. His path is carpeted with moss and lichen, with sedge and rush, with coarse red and yellow herbage, and with the rich green of the occasional bog patches which quake beneath his feet. As he clatters over a granite boulder, a mallard rises noisily, capping the rushes with his broad, strong wings, and sails swiftly away to a soggy fastness far overhead. Bearing in mind Colonel Hawker's advice, the sportsman halloos; whereupon a leash of Flappers flutter from the reedy margin of the stream, some thirty paces distant, shaping their flight in the course of the vanishing mallard. The gun is brought smartly to the shoulder. One of the three birds, a male, is a tempting mark, as he stretches his white-collared green neck above the high beech hedge across the water; and just as he is winging into top speed there is a puff and a flash, a bang reverberating along the valley, loudly bandied from hill to crag, and the bird is cut handsomely from sky to earth. He drops heavily into a placid pool, fully sixty yards away, whence he is brought by the carefully schooled dog. Before the rattle of the first discharge has ceased to re-echo, a second shot rings sharply out, and another Flapper is arrested in its flight, coming down headlong into a thorn-bush beside the stream. In fancy we see the sporting trio proceeding in single file upwards along the course of a tributary stream. In a hill-begirt amphitheatre they come upon a desolate fen, the stillness only broken by the lonely cry of the curlew wheeling overhead. The gunner stealthily approaches the leeward side of the marsh, picking his footsteps on the spongy margin of the swamp. Suddenly a plump of mallard dart from the reeds, and wing their rapid flight across the peat-red water. Promptly flashes forth a right and left in quick succession. There is a great fluttering of wings, and the dog, dashing in, quickly retrieves two well-favoured young duck.

In late autumn Flapper-shooting may be auspiciously prosecuted in regions near the coast, on estuaries and on small meres, either at daybreak or after sunset. Good sport is to be looked for by moonlight, when, after feeding, the birds bathe and preen themselves in some favourite watery haunt; though they are quite as wary in the bright moonbeams as in broadest daylight. Mallard and teal are well-nigh the only exclusively night-feeding water fowl. In the evening gloom, guided by the quacking, and by the sucking sound of many mandibles (not unlike the sighing of an ebb-tide),

FLAPPER SHOOTING.

they are to be killed at springs and fountains, and beside running waters. But it is not proposed here to go into the complicated mysteries of wild-fowling. The Flapper-shooter hunts the moorland turf-pits and green oases of bog patches; reed banks are worked and waded through; every available marsh walked; swamps beaten, and places at all likely to contain the succulent ducklings thoroughly investigated. Our game is not invariably hatched in close proximity to water. Occasionally a wild duck fixes on a most abnormal place wherein to construct her nest. The ten to sixteen greenish-white eggs have been found in the trunks and crowns of trees and in high and dry tussocks, away from either pond, pit or running water. Where the brood has been reared, there the Flappers for a time remain; and good shooting may be found in many a secluded bottom, and on many a moorland height. In such vicinage, and more particularly when alternately beating across wet areas and swampy hollows in the autumn, the pleasures of Flapper-shooting are likely to be increased by the flushing of a wisp of snipe, a spring of teal, or a company of sibillant widgeon. Much powder is burnt, the bag is heavy and varied, and the excitement of uncertainty is thrilling. Such chance encounters are things to be ruminated over afterwards by the ingle-nook.

Caution and sportsmanlike strategy will ever conduce to success in Flapper-shooting. The birds are shy, circumspect, and remarkably acute in sight and hearing. Though they delight to paddle and feed in fair, fresh water, such as lakes and streams, often somewhat exposed, their instinctive timidity is ever alert. Nevertheless, the good man, with good gun (held straight) and good dog, rarely fails to find the adolescent wild duck productive of pleasing sport and healthful recreation. That the birds require stopping when found and flushed is a fact incontrovertible. It has been estimated that a mallard's aerial speed is from forty-five to fifty miles an hour. Allowing an initial velocity of 1800 to 2000 feet per second for small shot, the sportsman must obviously hold well ahead in shooting at ducks on the wing. Flapper-shooting cannot be likened in simplicity to rat-hunting or to shelling peas. It is sufficiently difficult to be fascinating; its surroundings are usually romantic; it is worthy of the casual attention of the most expert and fastidious sportsman; and it is neither so dangerous as big game shooting, nor so arduous as the pursuit of the stag, the grouse, or even the partridge.

SEPTEMBER.

PARTRIDGE SHOOTING.

By Oswald Crawfurd.

No kind of sport has so changed with the times as Partridge shooting. Our forefathers shot them over dogs, and so far as sport is concerned, it is perhaps a pity that we have gone so far as we have in abandoning this old and delightful system of shooting; but, so far as the bag is concerned, the modern way is far the more profitable.

Why have things so changed, or, as some will have it, so degenerated? Some say it is because the French Partridge—the "Red-leg," introduced into this country from the Continent about 1750—has been mustering and breeding in countless multitudes ever since. The "Frenchman" runs before the dog, and ruins the staunchness of the best trained pointer, getting up out of shot—and, to a quicker tempered generation than our own, in the most damn-provoking manner conceivable; but shooting over dogs is obsolete in counties where the Red-leg has hardly penetrated. It can hardly be the Red-leg, then, that has done away with the dog. Perhaps the drill cultivation that allows the bird to run comfortably along a covered-way of turnip or mangold leaves for 30 or 40 yards before he rises, and so to puzzle the pointing dog, has helped in the innovation.

Probably, however, the fact that grain crops are no longer reaped by the sickle, but shorn by the scythe or the machine into stubbles that no longer can hide a titmouse, let alone a fat Partridge, is the chief cause of the doing away of dogs. The stubble being no longer tenantable by the covey, the cover area is reduced to clovers, colzas, lucerne and saint-foin, all broad-sown crops, or to such cover as bracken, gorse, heather or other rough wild growths, in which the Partridges now lie comparatively thicker on the ground than used to be the case, and the birds can be walked up by men in line without help of pointer or setter to quarter and search the fields, as was required when the area they lay over was greater.

So much for putting up of birds by "guns" and beaters without use of dogs, but late in the season, when partridges get strong on the wing and shyer of man, with more experience of that too common and dangerous predatory biped, this method of attack fails, and the birds would be left in peace till the ensuing autumn if no newer method of attack had been invented. If none had been invented three deplorable consequences would ensue. First, larders would be barer. Secondly, the older birds would accumulate, they being the most cunning, and the last to fall victims to the gun in the ordinary way of shooting; the old birds being peculiar in this, that when they are coupled for the season, they suffer no brother birds near the nest; consequently, the existence of many old birds on an estate means a small stock of Partridges in the ensuing season. Thirdly, the Red-leg Partridge would escape too frequently, and preponderate, were the shooting by lines of "guns" and beaters only, by his habit of running away unseen, whereas the gamer Grey Partridge flies up and is shot.

In 1845, Lord Huntingfield met these three difficulties by

instituting the system of driving. The guns are posted behind a suitable and stationary shelter, and the birds are driven over. The old birds, the wild ones and the red-legs, are the first to come over the posted shooters, and are the first victims. Driving gives worse sport than the old way, but it takes better shooting, for the birds come very fast overhead, and come by at every imaginable angle. Then again in driving all the generalship is displayed by the keeper with his army of beaters. The " gun " now need know nothing of the haunts and ways of birds. He is no longer a sportsman with all about him that the word sportsman conveys ; for the time being he has degenerated to a mere shooting-machine ; all he needs is a quick eye and straight powder.

The driving of partridges is mainly the sport of the rich man, armed with expensive shooting-irons, and waited on by a host of beaters, but partridges afford a vast deal of sport other than driving to poorer men of the unleisured classes.

The partridge is *par excellence* the game bird of England. Every man, almost every boy in the English counties who handles a gun at all, may hope to bring down his few brace in the autumn season. All through the spring and summer the "birds" and their breeding have formed the subject of rustic talk. "Wine talk is very pretty talk," said Thackeray, leniently, of the eternal "shop" indulged in by post-prandial man upon this important and little understood topic, but partridge talk, in summer time while the crops are growing thick, is far prettier talk. No one quite knows how many brace of old birds have been left from the year before, how many birds have paired, how the eggs have hatched out—and even where all goes well, and stoat and carrion crow and sparrow-hawk and polecat have spared the newly hatched chicks—a thunderstorm and a flooding rainfall may drown the "cheepers" over a whole county by the thousands. Then, when the beans and the clover grow up, and the corn crops are standing green and tall in the valleys and plains, in July and August, the birds show but little. Partridges run much at all seasons and fly only when they must, so that, with all the wealth of summer growth and their dinner table and beds, so to say, spread all round the birds, one hardly guesses how many and how strong the coveys may be.

The chances for the First of September are accordingly a most agreeable subject for conversation among rustic men; and the shepherd, who at daybreak has seen the partridges sunning themselves on the upland ridges, and, having put them up, has seen them wing their way quietly to the valley beneath him, counts them as they fly, and marks them down in the standing corn. "I seen a smartish lot on 'em"—no need to say partridges, or even "birds," as the First of September draws near—"I seen a smartish lot on 'em a-sunning theyselves on Clevedon Edge this morning;" and the simple remark causes a little flutter of interest in the village tap-room. "Did you happen mark 'em down, shepherd?" asks the gamekeeper, who has strolled in just to pick up some such scraps as this. If anyone of less social importance than the squire's keeper had asked this question, the shepherd would doubtless give an evasive answer, but to him he is constrained to tell the plain unvarnished truth. "Well, I'll tell 'ee true, keeper; they fled into Squire Joyce's wuts—that's where they fled, sir. I counted eighteen strongish birds." That is interesting hearing to everyone in the parish, and the odds are the keeper "stands a half-pint all round" on the strength of the news.

When the broad fields of oats and barley are cut, and the stacks are standing over the ground, as it mostly is before the end of August, the mystery of the coveys is at end; the partridge must fly, where before they only ran or cowered, and their whereabouts, their number and their size is common knowledge to everyone with eyes to see.

For those who pursue the popular sport of partridge-shooting in the simple, older fashion, with a brace of pointers and a retriever at heel—and no new-fangled method can give better sport, though, as I have said, it may give a better bag—there is no moment of the day so full of a delightful expectancy as when the first turnip or clover field is entered; the dogs are " quartering " well ahead; we see by their eagerness that they are near game; at any moment the "point" may come; then the sudden whirr of wings, the seeming confusion, the four or five gun reports, the dropping birds, and all the old, familiar incidents of the day—old and familiar, but never stale.

HUNTING THE WILD RED DEER.

By H. H. S. Pearse.

To understand the fascination that stag-hunting has for all classes in the Devon and Somerset country, and for a vast number of sportsmen from many distant corners of the earth, one must spend more than a few days of a single season on Exmoor. The charms of a sport that is pursued in bright August weather on broad stretches of moorland, eight hundred feet above sea level, or amid the deep shadows of wooded valleys musical with brooks, all lovers of nature may, in a general sense, appreciate. Some fox-hunters, and especially Meltonians, who estimate all sport by one standard, are apt to regard as slow a method of pursuit that lacks the first wild reckless rush in which their ardent spirits revel when a " Gone away !" is sounded with the Quorn or Pytchley. Those who do not stay in " Red deer land " long enough to let that impression wear off, will perhaps carry away a belief that Devon and Somerset men in their simplicity are content with very little. But anyone who has seen a great run over these moors will want to see another, and when he has learned something of the woodcraft which the harbourer and huntsman can teach him, he will be in a fair way to comprehend the enthusiasm of men who declare that there is no sport in all the world that can rival the chase of the wild red deer. A stranger desirous of knowing all its delights had better not begin stag-hunting in the early August days. During the first two or three weeks of each season only old harts, heavy with the fat of summer idleness and too cunning to run straight, are hunted. The harbourer is proud to show his skill in slotting the heaviest deer of a herd, and a master who knows his business will have these killed before he allows the five or six-year-old gallopers to be pursued. In the interests of farmers, who are the best friends of stag-hunting, it is important that cunning veterans of the forest should be sacrificed first, for they do more damage to crops than twice as many younger ones.

These old harts, however, are not brought to bay without the exercise of much skilful woodcraft, and, though such hunting may lack the element of rapturous excitement, it has charms for all who love hounds and understand their work. Let the stranger with these qualifications get permission to accompany Anthony Huxtable into covert when the tufters are taken to where a heavy stag has been harboured. Three or four couples of the oldest and most trustworthy hounds are selected for this work, and the remainder of the pack kenneled in a barn, or any building that may be handy. To throw all the hounds into covert at once would be to defeat the first object of tufting, which is to rouse the harboured deer and no other, though a score or more may be lying in the thickets of the same great woodland. The stranger should provide himself with a sure-footed hack for this work, leaving his hunter where the pack is. The tufters will perhaps have to thread that wooded valley again and again before they can force the right stag away, and one who would watch them at every turn must be prepared to gallop up or down rough paths, steep in places as the roof of a house. A deep-mouthed chorus echoing from crag to crag down the valley, followed by a shrill, wild holloa that makes one's nerves tingle with rapture, proclaims that the old stag is on foot, and the fox-hunter listens for the "gone away," or dashes eagerly forward in his anxiety not to be left behind. But the next sound he hears is a whipper-in rating the tufters with "get away back to him!" "ware hind!" and, obedient to a peculiar blast of the huntsman's horn, these clever old hounds will swing round to where their proper game was last seen. The wily stag is a master of every shift and subterfuge. He doubles on his tracks like a hare, beating up and down the covert, turning out hinds and young male deer one after another, and lying down in their lairs until he is fresh found. All this may be the work of hours, but it is fruitful in lessons of such scientific hunting as one cannot see elsewhere. The line, when lost, can only be recovered again at times by the exercise of a skill and keenness of perception that rival the Red Indian's faculties.

At last the welcome "Tally Ho!" is heard on the open moor, and one may be sure now that the right stag has broken covert. The whipper-in, or some trusty follower posted there, has seen the monarch of the glen crash through the oak copse, pause a moment to sniff the air, turn his beamed frontlet so that every "right"— brow bay and tray, and three on the top—may be counted, and then with a defiant stamp of his forefoot bound off across the heather. Instead of galloping hurriedly towards that point, the huntsman rides back blowing his horn as a signal, while the whipper-in goes forward at speed to stop the tufters. Now there is mounting in hot haste at the farm where old stag-hunters have

HUNTING THE WILD RED DEER.

been possessing their souls in patience so long. The hounds are set free, and the huntsman takes them off by paths he knows well to the distant moor. By the time they reach the spot where Jack stopped the tufters, twenty minutes, half-an-hour, possibly an hour may have passed since the deer went away. But they feather on the line at once, own to it with a whimper that swells into glorious melody as they feel the "titillating joy" of sweet scent that clings to the heather; and now you may ride, for they will take some catching if you are far behind them when they breast the next hill. Men who have been long at this game do not try to go straight down and up the steeps, where loose "shillets" clatter under hoofs at every stride, if they can get round more quickly and easily by skirting the Coombe. Following hounds over such a country, with its alternations of deep valleys, rugged ravines, and soft, if not boggy ridges, is an art that can only be acquired by practice, as many a Leicestershire man has found to his cost after riding his horse to a standstill in vain endeavour to live with hounds that seem to go so slowly. The pack will stick to their hunted deer, though he may run through almost interminable woodlands haunted by other herds. Though often at fault, these hounds turn quickly to every note of their huntsmen's horn, and puzzle out the tangled thread of scent again and again with wonderful sagacity, so that a stag once found rarely shakes them off. If he do not take refuge in the sea they will " set him up " before nightfall in some shadowy pool with his back to a rock, where he must fight for his life. And he does fight gallantly, with no trace of fear for the foes that clamour fiercely round him. When he takes to the sea by Bossington or Porlock or beautiful Glenthorn he swims so well that no hounds would ever overtake him if boats were not at hand to

F f

effect a capture. More than once a hunted stag has been known to cross the Severn Sea, and land in safety on the shores of Wales—a safety which no true sportsman would grudge him. When the hunting moon has waxed and waned, when stags have gained their lustiest strength, and the mists of chill October give token that the season of " love and war " is at hand, a six-year-old deer will lead pursuers many a league over the moors, and many good steeds will be sobbing before their riders see the noble quarry brought to bay in the dark pools of Waters' Meet or under the wooded banks of the Barle, or where the swift stream tumbles among giant ferns and grey boulders in Horner Valley.

RABBIT HAWKING.

By J. E. Harting.

Hawking, like other field sports, has its seasons, and just as there are various breeds of hounds to suit the nature of the different animals hunted, so are there different kinds of hawks according to the nature of the "quarry" at which they are flown.

For rooks in the early spring, and for grouse and partridges in autumn, the peregrine falcon is used; for larks in August, a cast or merlins is employed; the sparrow hawk shows good sport with blackbirds and thrushes in the turnip fields, and was formerly used for taking landrails and quails; while for ground game, as well as for an occasional partridge, pheasant, or moorhen, the Goshawk is without a rival. This is especially the case if the falconer lives in a woodland or enclosed country, where, from the peculiar nature of its flight, a long-winged falcon is liable to be frequently lost, since it rises to a considerable height, ranges wide, and stoops at the "quarry" from a great distance, often killing it out of sight of the falconer, should a copse or other covert intervene. For this reason a long-winged falcon like the peregrine should be flown in a very open country, such as a wide, flat moor, or cultivated downland, where there are no trees, and where hedges are few and far between.

The Goshawk, on the other hand, with a different mode of flight—going straight from its owner's fist to the quarry at which it is flown—is essentially the hawk for an enclosed country, and if properly trained in the first instance should never be lost. It is a mistake, however, to suppose that anyone inclined to the sport, though knowing nothing of it, may purchase a trained Goshawk, take her out the next day, and fly her without any risk of losing her. She would almost certainly be lost for want of the requisite knowledge on the part of her new owner how to manage her. You may buy a new horse or a new dog, and hunt the one or shoot over the other possibly without disappointment; but it is otherwise with a hawk. Hawking requires an apprenticeship, and no one can expect success who has not gone through the various stages of taming and training his own birds, spoiling some and losing others, until he has discovered his mistakes by dire experience.

There was a time, before the art of shooting flying came into vogue, when almost every country gentleman in England kept a Goshawk or two, and very high prices were given for well-trained birds. Even in James I.'s time, after "birding-pieces" had been introduced, good Goshawks fetched a good round sum. Edmund Bert, who published "An Approved Treatise of Hawks and Hawking," in 1619, tells us that he had "for a Goshawke and Tarsell a hundred marks, both solde to one man within sixteen months," and for another he was offered forty pounds, and ultimately sold her for thirty—an extraordinary price, when we consider the relative value of money in those days. At that time,

RABBIT HAWKING.

doubtless there were many places in the British Islands where
the Goshawk was to be found breeding, where the nests, or eyries,
were jealously watched, and the young were taken as soon as they
were ready to fly. This, however, is a thing of the past. It is
very many years since a Goshawk's nest was found in Great
Britain; not since Colonel Thornton, a well-known falconer and
good all-round sportsman, discovered one in the Forest of
Rothiemurcus, and trained one of the young birds. This was at
the end of the last or beginning of the present century, since
which time no similar discovery has been recorded.

At the present day, the Goshawks trained and flown in England
(and we know of many) are procured from France or Germany;
chiefly from France, where, thanks to the good offices of some of
the French falconers, they are looked after, the nests protected,
and the young birds secured at the proper time. The price
varies with the age and condition of the bird. You may get one
through a dealer for a couple of pounds, but it is a chance
whether the flight feathers are unbroken, perfect wings being a
sine quâ non in the case of a hawk that is to be trained and flown.
It is better to pay a little more, as in Paris, and secure a good
one. Occasionally a Goshawk is taken in a bow-net by one of the
Dutch hawk-catchers in North Brabant, and sent to England;
but as a rule the birds captured by them are peregrines, for
which, at the present day, there is greater demand.

As to the mode of training, it is very simple when you know it,
and the falconer who gives the greatest amount of personal
attention to the matter will have the greatest measure of success.
A hawk must learn to know her owner, or she will not allow him
to take her up. She must be fed by him, and carried as much

as possible on the glove, bare-headed, that is, not hooded, to accustom her to the sight of men and dogs, and other moving objects, that she may put off all fear, and become as tractable as any pointer or setter; knowing her owner's voice, and obeying his call or lure. The old falconers used to say that a hawk should know no perch but her owner's fist, and there is a good deal of reason in this, for the more a hawk is carried the better she will be.

The first step is to induce the bird to come off the perch on to the glove, which is always worn on the left hand, to leave the right hand free for detaching leash and swivel before she is flown. This end is usually attained by offering a little bit of meat in the glove, or the leg or wing of a fowl or pigeon; but only a mouthful should be given as a reward for obedience, for it must be remembered that a hawk must be flown fasting, and rewarded for killing, or for coming back after an unsuccessful flight; and if too much be given at a time, her hunger is thereby appeased, and she has no longer any incentive to hunt. When she will step readily off the perch on to the fist, the leash being untied, the distance should be increased from a foot to a yard, and at length to several yards, until eventually she will fly willingly across the room to her master. This lesson being repeated out of doors, from a field gate or the top of a stone wall, while for safety a long line is tied to the swivel—which in turn is attached to the jesses, or little leather strap round her legs— in a few days she will come readily on being called, and the line may then be dispensed with. She may then be lured with a dead rabbit, or part of one, thrown down and drawn with a line along the ground. After coming readily to this several times, she is next to be entered to the live quarry. For this purpose a young rabbit or two may be easily procured by ferreting, and being placed under an inverted flower-pot which can be pulled over from a distance, with a piece of string and cross-stick through the hole in the bottom, the hawk is slipped at the right moment, and rarely fails to take the rabbit at the first attempt.

Another trial or two of this kind, and she is ready to fly at a wild one. The critical part of the training is now at hand, and great care must be taken to avoid disappointing the hawk ; that is to say, the rabbit should be well in the open, and not within easy reach of a burrow into which it may pop just as the hawk is about to seize it. Encouraged by the success of these first attempts, she will go on improving every day, and the more she is carried and flown the better she will become.

To show what success may be attained even in the first season with a young Goshawk, we may refer to the bag made by a falconer still living, who in his first season, with a young female Goshawk (better than a male bird, because larger and stronger) which he trained himself, took 322 rabbits, three hares and two magpies, and the following season 280 rabbits, two leverets, eleven partridges, four magpies and two squirrels !

After this no tyro need despair, and though, for want of experience, he may not attain to such success as this, he will at all events discover in the sport of rabbit-hawking a most fascinating and enjoyable recreation.

The accompanying illustration shows the way in which a Goshawk leaves the fist when the quarry is found and started. It is not to be supposed that there is any unnecessary cruelty in the sport, for the falconer only teaches his hawk to do for his amusement what she has to do every day, when in a wild state, for her own living; and the accomplishment of this affords one of the best and most curious illustrations of the extent of man's power over the lower animals.

OCTOBER.

PHEASANT SHOOTING.

By George Lindesay.

Originally a native of Asia Minor, the Common Pheasant is a wonderful instance of the successful acclimatization of a foreign bird in these Islands, and a most important inhabitant thereof has he become. Hundreds of thousands of pounds are invested in the preservation and multiplying of his species; no small portion of rural England is reserved for his residence; and up to a certain date his person is as sacred as that of a fox. After the 1st of October his importance increases; and thousands of sportsmen, armed with the latest and most quick-firing of breechloaders, accompanied by keepers and beaters innumerable, march for his destruction; they violate with their hob-nailed boots and gaiters the coverts that have hitherto formed his sure sanctuary, and poor *Phasianus Colchicus* appears suddenly and simultaneously in every poulterer's shop in the United Kingdom.

Some idea may be formed of the wonderful increase in the number of pheasants which has taken place in this country, from the figures given in the Badminton Library volume on shooting, as applying to some 10,000 acres of preserved land in Norfolk. On this estate in 1825 the total bag was 89 pheasants to one gun; in

1860 it had increased progressively to 2256; and in 1881 there were shot no fewer than 5363 pheasants, the best day being 1135 head to eight guns. Of course this enormous increase is due to artificial breeding and strict preserving — for the pheasant, especially in his early youth, requires much shelter and plenty of food—but it shows how much these may do when carefully carried out, materially benefiting the proprietor and the food supply of the community as well as the sportsman. The young birds are mostly fed upon ant eggs, maggots, and grits, but when grown they eat seeds, roots, and leaves indiscriminately. The cock is far from being a specimen of domestic virtue; he takes no notice of his offspring, and during the greater part of the year he leads a very independent kind of existence, associating mostly with others of his sex, while he frequently mates with the common hen, and sometimes with the grouse, turkey, and guinea-fowl. He is also of a very pugnacious disposition, and often enters into mortal combat with the barn-door cock, over whom he has no small advantage owing to his powers of flight, which enable him, when fatigued, to ascend into a tree for recuperative purposes. Occasionally the hen-pheasant chooses the deserted tree nest of an owl or squirrel wherein to lay her eggs, but her ordinary nest is a very rude construction indeed of leaves and grass, placed in a slight depression of the ground, with hardly any attempt at concealment; the eggs are of a uniform olive-brown colour, and usually from eight to ten in number.

The Pheasant is an undoubted friend of the farmer, for he destroys vast numbers of injurious insects—a form of food which he loves beyond others—and over 1200 wire-worms, that worst pest of the farmer, have been taken out of the crop of a single

bird. During the day he trusts more to his legs than his wings, but his habit of roosting among the trees makes him a conspicuous object, and an easy prey to the poacher on moonlight nights. The ingenious dodge devised by Mr. Waterton for the deception of these gentry, and the preservation of his birds, by having wooden pheasants nailed on to the boughs of his trees, is well known.

Among other varieties, the Golden, Silver, and Reed's Pheasants have been also introduced, with more or less success, on some of the great sporting estates in England. The last-named is, perhaps, the handsomest bird of his handsome race; the body no larger than that of the Common Pheasant, a cock in full plumage yet measures 8 feet from head to tip of the tail-feathers, which are themselves 6 or 7 feet in length, and beautifully marked. Reeve's Pheasant, or, as it is called in France, *le faisan révère*, will more readily take to wing than most of his kind, and although so splendid and highly-coloured, is a hardy bird, his home being among the snow-clad mountains of Surinagar and in Northern China.

Both the Golden and the Silver Pheasants are also natives of the last-named country, in which the former especially is held in great esteem, not only for his elegance of form and splendid plumage, but also for his delicacy of flavour, which surpasses that of the Common variety. It is hardly necessary to remark upon the value of the plumage of this gorgeous and very beautiful bird to the salmon fly-fisher, who greatly prizes his crest, neck, and tail-feathers.

The Silver Pheasant is a much larger and more powerful bird, and like his Golden brother has become an inhabitant of a good many British preserves; his weight, however, being out of proportion to his strength of wing, is rather against him, and he falls an easy victim to stoats, foxes, and polecats.

Battue shooting, of course, accounts for the vast majority of pheasants which are killed in this country, and on this subject innumerable volumes have been written, and a good deal of acrimonious talk indulged in. Bromley Davenport criticizes in amusing language the fallacies put forward by many of those who speak and write against battue-shooting, and describes the great amount of skill and management required to ensure success. The fact is that this form of shooting has its own special attractions, but can only be indulged in by the rich, and is on that account much cavilled at. To bring about a big and successful battue, large sums of money must, in the first place, be expended in breeding and rearing the young birds, and in preserving them; the responsible manager of the drive, too, the man whose duty it is to bring the birds together, and make them fly over the line of guns, as arranged and distributed beforehand, has no easy task. He must be thoroughly acquainted not only with the ways and habits of the birds, under all conditions and circumstances, but he must be alert to recognize the modifications to which these are subject, from wind, weather, or other temporary influences. As I have above remarked, the Pheasant is much fonder of running than of flying, and a not unimportant part of the head-keeper's duty is so to manage that the birds are brought forward as much as possible by means of their legs, and that they do not take wing in any considerable numbers until brought into contact with the line of guns. As a rule, it is preferable to begin by driving in the birds from the most outlying coverts and woods, the fear of scaring them across the march being generally exaggerated, as when driven from the ground they are accustomed to frequent, pheasants will always endeavour to return thither. Another essential to success in covert-shooting is the judicious placing of stoppers; late in the season more especially, when the birds have probably been disturbed, and are wild and unsettled, these are even more important than beaters, and should be placed at angles of coverts, or where hedge-rows and the like impinge upon the main wood, else numerous birds will steal away and escape.

Personally, I confess to a preference for a form of sport less artificial, and one in which the skill and knowledge required are more centred in the shooter himself than in his servants. A high, rocketing pheasant at a battue is doubtless not such an easy object to bring down as would appear, and offers a more difficult shot than one which gets up a few yards in front of a setter's nose on the edge of a turnip-field or bit of open covert; but in my humble opinion there is no getting away from the fact that there is more of sport about the latter performance, and that a day with the keeper and the dogs along the hedge-rows and thin strips of outlying covert is more enjoyable than a hot corner in the home preserves.

One of the most pleasant covert-shoots I can recollect was that in which I used to take part year after year on the banks of Tweed. My host, the laird, had a fad in regard to his pheasants, and, except outlying bits of wood, never touched his best coverts until the last week of the season; then we shot them every day until the 31st, and began salmon-fishing the very next, the opening day, the 1st of February. At that time of the year, however, fishing on Tweed was often a precarious pastime, a clean fish being a great rarity and the water full of kelts. It was a singular, and not very pleasant, sight to see bodies of the spent fish lying in hundreds among the ice-blocks and snow on the banks, left there by the rapid subsidences of the river. Great salmon, 20, 30 and 40 lbs. weight, were there lying dead, the crows and other birds feeding upon their decomposing bodies. Though the salmon might fail us, however, the laird's pheasant-shooting was excellent ; the birds, strong on the wing and full-grown, rendered straight holding and choke-bores absolute necessities ; a fair number of woodcock was always a certainty, while the way the bunnies lay out was astonishing.

I can only recall one day on which our sport was absolutely stopped through stress of weather, and it was not until some days afterwards that we ascertained we had attempted shooting in what proved to be a quite historical gale. By 11 a.m. the wind had increased to hurricane force, and, driven before it, blinding clouds of snow continued to descend without intermission. The beaters had entered one fine piece of covert before things had reached this climax, but the drive was a complete failure ; most of the birds simply could not get up, and those that did manage to rise a yard or two from the ground were whisked away in an instant by the tremendous gale. We were truly thankful to make our way home to the shelter of the old mansion house, from whose windows we were enabled to gaze with equanimity upon the snow, as it wildly eddied and whirled before the storm.

CUB-HUNTING.

By H. H. S. PEARSE.

IN counties where great chains of woodland alternate with open downs or wide heathery wastes, broken only by a few cultivated patches, Cub-hunting may begin in the sultry days of August. No crops are there to be injured, and the number of sportsmen who gather in the dewy morning to meet hounds at such places is never great enough to do much damage in any case. Some fashionable hunts are equally favoured in another way, having vast coverts through which hounds may work for hours, day after day, and rattle the cubs about without risk of forcing one out across an acre of corn.

At Badminton, the Marquis of Worcester begins cubbing the first or second week in August, and finds within the confines of that noble park sport enough to occupy him until the harvest has been gathered, and every puppy in the pack has learned what the duty of a well-bred foxhound is. Even in such a pack—where hereditary instincts of the highest order are transmitted according to scientific rules—and under such tuition the young ones do not all learn their business with equal readiness. There, as elsewhere, one finds the over-impetuous, that will hunt anything, until they have been broken of their tendency to run riot, and the slack that show an inclination to hunt nothing at the outset. Before the frosts of chill October begin to bring down the sapless leaves, however, they must learn all about it, or the Badminton Kennels will know them no more. In circumstances like these, with ample time for training, and almost unlimited resources to draw upon, it is not surprising that the Marquis of Worcester has a pack from which the wiliest and stoutest foxes find it difficult to escape. I know of no hunt in England where the process of teaching cubs to run, and young hounds to hunt them, can be more profitably studied than at Badminton. In the broad green rides of those well-stocked coverts one may enjoy the freshness of August mornings, while one watches the hounds working patiently among the dense undergrowth, and listens to their music as it swells from an uncertain treble into a clamorous chorus. A practised ear can detect every change of the chase. Some shrill notes, but low and tremulous, tell that a puppy has something to say, but has not confidence enough in himself to give tongue boldly. If these faint whimperings are followed by deeper notes, one may be sure the young one was right, and the hounds know it too, for they come together with a crash that shakes the copse. But cubs are many, and in their confused efforts to escape, they dodge about in a way that is more puzzling than all the tricks of which an old fox is master. Sudden silence shows that the hounds are at fault, and when a tongue one has not heard before rings out clearly, it is a sure sign of a fresh scent.

Then, perhaps, the puppies will speak together with the temerity of impetuous youth, while their elders look at them with a silent reproof that should be more cutting than the whipper-in's stern "ware riot," to a young hound of proper feeling. As the hounds know each other—which to put faith in implicitly, which to mistrust as a forward youth who speaks before he is perfectly sure of being right, and which must be watched closely lest he steal away in jealous silence without giving a signal by so much as one low whimper—so the huntsman gets to know them and all their peculiarities in the months of intimate association between August and November. Thus, and thus only, can a pack be raised to such all-round perfection as the Badminton has long been celebrated for.

Most hunts have to be content with a more rapid and much less complete process of training for the work of each season. They may be divided broadly into two classes, one having such vast extent of woodlands that a run in the open is of rare occurrence, and the other coverts so small and far apart that, even in the early days of Cub-hunting, hounds rarely kill without having some sort of scurry across country. Woodland hunting pure and simple does not suit the taste of everybody. It demands special qualities both in the hounds and huntsman, while for full enjoyment of it, followers need be deeply versed in all the mysteries of woodcraft. The best huntsman in woodlands that I ever knew seemed to trouble himself very little whether his hounds stuck to one cub or changed to a fresh one, so long as they kept well together; but tendency to riot he would check with a note like a thunderclap, and that was more effective than any amount of thong the whipper-in could apply. If they kept changing, so much the worse for them. Their reward was all the longer in being earned, though perhaps the sweeter on that account when it did come. At all events the system worked well with him. A slack hound was never seen in his pack, which, before the regular season opened, would drive a fox from end to end of the vast coverts, and seldom leave that line for a fresh scent. Perhaps he had secret methods too deep for casual observers to fathom. The sort of sport, however, that most men can appreciate without professed study of scientific woodcraft is to be found in countries where coverts are comparatively small, and separated by open tracts not too formidably fenced. These need not be all pleasant pastures, as in the fashionable shires. A due proportion of freshly-turned fallow or dry stubble gives hounds a chance of showing their hunting powers, and prevents them from rolling the cubs over too soon. There, however, the sport must be deferred until nearly every acre of corn has been cut, and by that time the cubs are strong enough to take care of themselves, though the chances are that by being left so long undisturbed they have acquired stay-at-home habits, and do not know their way about the country. To teach them and the young hounds all that must be learned in little more than a month is no light labour for a huntsman, whose troubles are not lessened by the fact that chances of a gallop, however short, are certain to attract a number of impetuous youths for whom the slower incidents of woodland hunting in August have few charms. When the first light frosts of late September redden the copses, and misty sunlight rests on broad stretches of stubble, Cub-hunting may begin in the least-favoured countries. Then master and huntsman can shake off many cares concerning probable claims for damage to crops, but they will be still harassed by dread of the thoughtless pursuer, whose eagerness

so often spoils sport at a time when every lesson is of importance. The necessity for such training is not confined to young hounds and cubs. There are some men who hunt, year after year, without learning anything that the events of a single day at this season should suffice to teach them, and a knowledge of which would enhance their own pleasure, while contributing materially to the pleasure of others, instead of marring it. If they would only remember what an unpardonable sin it is to holloa one cub away while the hounds are running another in covert, and that they should not shriek wildly, but simply content themselves with a "Tally-ho! over," just loud enough for the huntsman to hear, when they view the hunted cub across a ride, much waste of strong language would thereby be spared. They had better hold their peace altogether if not perfectly sure that it is the hunted cub; and, at any rate, all shrill holloaing should be left to the whipper-in, whose voice the hounds know and can trust. There are always experts enough in the field to help a huntsman by timely tidings when they are sure he is at fault. Those who have not experience to guide them will be wise if they keep their mouths shut and their eyes open, watching every turn the hounds take as they work hither and thither among the tangled brambles and tawny bracken. The time for allowing hounds to get away will be hastened rather than retarded by a little patience at the outset. Before the bleak gales of October have blown many of the dead leaves down, the pack should have brought the weaklier cubs in nearly every corner of a hunting country to hand. A late beginning, where the coverts are neither very big nor close together, is not wholly disadvantageous. After the corn has been gathered lost days may be made up for, and by this time the cubs are stout enough to show sport of the kind for which impetuous youth has been fretting since the earliest notes of horn and hound were heard. The pack, too, knows its work now, or never will, and the master is nothing loth to let the mettle of his hounds be tried in chase of a stout-hearted rover. The joy that thrills young nerves when the whipper-in's halloa, no longer suppressed, tells that a good cub has gone away, is contagious. The oldest among us cannot resist it. He would fain gallop as they do when hounds are speeding over the rounded uplands; feel the keen rush of air that comes whistling out of the gloomy clouds, and enjoy the rapture of rivalry once more. One can hardly find it in his heart to blame even the man who, in his eagerness to be with the pack, crashes over a fence and rides headlong for the hounds, regardless of their having checked on the river bank, where he is in danger of killing them and drowning himself. And what a disappointment it is to the huntsman if he cannot write, "killed in the open" at the end of his hunting notes for that day. It is, no doubt, a laudable ambition for one who believes that the first duty, if not the sole mission in life, of hounds and huntsmen is to kill foxes, and we all do our best to encourage it while the run lasts. But somehow, the final scene that fills him with a fierce joy brings a touch of regret to some true sportsmen, who would always wish that a good cub might live to grow into a good fox, and to lead us many more glorious chases than any we are likely to see before the season of Cub-hunting ends.

PARTRIDGE HAWKING.

By J. E. Harting.

WE have described on a previous page the art of rabbit hawking with the goshawk. The goshawk is one of the short-winged hawks—so called because the tail exceeds the wings in length.

The *modus operandi* with a long-winged falcon, like the Peregrine, which is used for partridges and grouse, is very different, and depends upon the different manner in which these two hawks take their prey. A goshawk, as has been shown, flies straight after the quarry from the fist, and overtakes it by superiority of speed; a falcon, soaring in the air, captures its prey by descending rapidly upon it from a height, and this descent with half-closed wings is technically termed a "stoop."

To witness the stoop of a well-trained falcon at a fast-flying partridge or grouse, as it goes down wind at the highest speed of which it is capable, is a sight to be for ever remembered. Sportsmen who know what it is to shoot driven birds, and who are wont to estimate the speed of a partridge at forty, fifty or even sixty miles an hour, incline to the belief that a driven partridge must be the fastest bird that flies, and few are prepared to learn that neither a partridge nor a grouse can live long before a Peregrine Falcon. There is a popular notion and a pretty theory that hawks are " Nature's police ; " that they carry off the weakly birds of a covey, and so, by allowing the strongest and best to escape, they help to maintain a good healthy stock of game. This is an absolute fallacy. Having seen scores of grouse and partridges taken by trained falcons, we are in a position to assert positively that the power of wing in a Peregrine is so great, that it can overtake and strike down the strongest and best bird in a covey with as much ease as if it were the youngest and weakliest of the brood. It would naturally be supposed that on the rising of a covey, the hawk would stoop at the bird nearest to it. This is not invariably the case. We have many times seen a falcon ignore a grouse immediately under her, and single out for capture either the leader of the pack, or an outside bird far ahead of some of its fellows. It is the same in partridge hawking. The hawk probably stoops at the bird of which it first catches sight when the covey rises ; just as in partridge shooting the sportsman singles out the bird he first sees on the wing, unless it happens to be too near him, when he aims at one further away, by which time the bird first seen will be at a proper distance for his second barrel.

It will be inferred from what has been already stated that the modes of flying a short-winged hawk and a long-winged falcon are entirely different. In the former case the quarry is found before the hawk leaves the fist, in the latter the falcon is put on the wing and allowed to " mount " to a good " pitch " before the dogs are uncoupled and allowed to range. As soon as they are steady on point, and the falcon well placed overhead, the birds are flushed, and the falcon, immediately catching sight of them, with a headlong rush, stoops at the one she has singled out. So true a judge is she of pace and distance, that, unless the partridge drops suddenly into covert, she rarely fails to strike it fatally. No prettier picture could present itself to the eye of an artist than the grouping of dogs, hawk, and falcōnēr at the moment which precedes the fatal stoop. The dogs—setters or pointers, as the case may be—motionless on point, half-concealed perhaps in a field of turnips, or patch of clover, or standing out in bold relief upon the edge of a stubble; the hawk, well understanding the proceedings, " waiting on " at a considerable height above them; the falconers, advancing slowly in line, or pausing in their enthusiasm, to admire the scene before them. A step forward, a rush of wings, a shout of " Hoo, ha, ha," a grey meteor falls across the sky, and amid a small cloud of feathers a partridge drops with a dull thud amongst the turnip leaves, or disappears like a stone in water, in the waving clover.

A finale such as this, however, is not to be effected as a matter of course by any tyro who can procure a hawk. Its accomplishment implies a good deal of previous trouble in the taming, training, feeding, bathing, and general management of the noble falcon before it can be trusted to fly at liberty, and exhibit the exercise of its natural instinct for man's pleasure and benefit.

It would be impossible within our present limits to give anything like a detailed account of the mode of training gamehawks, a subject upon which many books have been written; but, with a view to encourage some attempts on the part of those who have the leisure and inclination for such sport, it will not be out of place to offer a few remarks upon the more important points to be attended to. It may be stated, then, as a general principle which underlies the whole art of falconry, that a hawk is flown fasting, and is rewarded for killing, or for coming back after an unsuccessful flight. Hence the use of the "lure"— a dead pigeon at the end of a string, or a couple of wings tied together and weighted, and garnished with some raw meat, which is only shown to the hawk at feeding-time, or when she is required to return to her owner, or, again, if she is too far down wind, when the dogs are "standing." As a rule hawks are fed but once a day, about five o'clock in the afternoon; but Merlins are all the better for having a light morning meal in addition, about 7 a.m. Indeed, we have found it a good plan to give all hawks a mouthful or two in the morning, after they have got rid of their casting (that is, after they have thrown up the indigestible portion of their food, in the form of an oval pellet), and at the moment of taking them from the "perch" to set them down upon the "block" to bathe. This puts them in good humour, prevents them from "bating" too much, and an hour or two after they have got perfectly dry they are keen and in good order for flying. After a hawk has been "called off" to the "lure," at first with a "creance" or long light line attached to the "jesses," and afterwards without it, she has to be "entered" to the particular "quarry" at which she is intended to be flown. This can best be done by previously shooting a partridge, and, while the hawk is on the wing, throwing it out to her in a long line with which she can be checked in case she should attempt to carry the bird away.

When this has been done a few times, the hawk being always allowed time to break into the "quarry" and get a few good mouthfuls before being taken up, she may be flown at a live partridge. And

PARTRIDGE HAWKING.

here it should be noted that it is all-important not to disappoint the hawk in her first flight. To avoid this it is a good plan not to unhood her and put her upon the wing until a covey has been found and marked down. The hawk may then be flown, and the falconer, walking towards the spot where the birds have "put in," will be careful not to flush them until he sees that the hawk is well-placed and with her head towards them, so that she may see them the moment they rise. By this plan he will ensure the best chance of success; for if a hawk kills the first time she is flown, it will be the making of her. Another piece of advice we would give is, never to run up to a hawk the moment she has killed, but give her time to plume the quarry and break into it, approaching her quietly, and, when near enough, kneeling down with a bit of meat or a partridge wing in the glove, and holding it under her. She will at once seize it, and, stepping on to the glove, may be lifted up gently by aid of the jesses, which must then be firmly held, lest she should attempt to fly. The mischief of "making in" too quickly to a hawk is that it alarms her, and causes her either to carry off the bird she has killed to a distance, or to fly away without it and give some trouble before she is retaken. The fault of "carrying," thus induced by want of care on the part of the falconer, is one that by all means should be guarded against from the beginning.

On taking a hawk up from the quarry she should have a mouthful or two given her by way of reward, and it is not a bad plan to pull off the head of the partridge—or grouse, as the case may be—and, crushing it to pieces, allow her to eat the brain and such tit-bits as she can get off it. The hood being then replaced, she is ready to rest a while before essaying another flight.

Above all things, gentleness with a hawk is a *sine quâ non*, and a light hand in hooding. The bird has then nothing to be afraid of. Instead of being alarmed at the approach of her owner, lest he should rob her of her prey, she comes to regard him as useful in helping her to secure it, allows herself to be lifted up on his glove, feeds before him, and exhibits every sign of confidence and affection. Indeed, it is not too much to say that a hawk that has been properly handled from the first by a kind master will become as obedient and as much attached to him as a favourite dog. The wonder is there are not more falconers!

ROUGH SHOOTING.

By H. A. S. Pearse.

The man who can rent a thousand acres of rough shooting ground on the borders of a certain moorland I know of has material for health and contentment within his reach. He need not envy pheasant preservers their big battues, nor deer-stalkers their wide domains. If it be no more than a mere strip averaging five hundred yards wide and following the tortuous course of a river from rush-grown moor to wooded valley, so much the better for chances of sport within its limits. With breechloader in hand and a brace of spaniels for companions, one may wander about the ridges and hollows of such ground week after week in proper season without exhausting its resources of feather and fur, or finding the frequent repetitions of familiar scenes monotonous. A keen observer of nature sees endless variety within the narrowest limits. Now it is the colour of foliage and ferns that changes, or the form of trees as they cast off the gold-embroidered vesture of regal autumn, or the river churned by winter floods into foam that whitens tawny pools whereon, in rich mosaic, leaves, flowers, sky, and sunlight cast their clear reflections but a few months ago; and now the notes of animate nature making fresh harmonies for every day. Only an ardent lover of nature and keen sportsman can appreciate such things fully, but it is he alone who cares for rough shooting in preference to all other, or knows the delight of trudging from morn to eve through tangled heather and tall brake ferns, by reedy swamp and rugged coombe, for the sake of a bag that pot-hunters would consider beneath contempt. It is not the weight so much as the variety that gratifies him; and every bird, rabbit, or hare bagged may represent in his eyes a triumph of skill in shooting or in woodcraft. He does not assume superiority over nor affect contempt for the battue men, as some of us do. He acknowledges readily their dexterity in bringing down a "rocketer" or stopping the leader of a driven grouse pack as it skims past a peat stack; but he thinks—not without reason—that they lose more than half the sport in not finding their own game. His great pleasure is to watch well-trained dogs at work, or to exercise his own skill when their sagacity fails.

Among the trophies that he displays with pardonable pride, you may perhaps see a black cock or grey hen, whose wary wildness was only out-matched by patient strategy, of which few but moormen know the alphabet. Black game are not difficult to get within range of during the sultry August or September days, when they are fat with good feeding; but with the first frosts of autumn they become so shy that they are more difficult to stalk than a cunning stag, and nobody who does not know all their habits can drive them to the guns. They seem to scent powder, too, a mile off, and though a shepherd may walk close by them without disturbing one, the faintest sound of a sportsman's footsteps brushing the heather will cause wily old birds to rise on whirring wings and dart away long before he can get within range of them. The arts by which a man accustomed to rough shooting from boyhood will out-manœuvre the wildest of wild birds are innumerable, but he cannot perhaps practise the same trick twice running with equal success. Curlews I have known so shy and cautiously clever that they would lead a man on from hill to hill without giving him the chance of a single shot at them ; and, after all, I have succeeded in getting close enough to kill with each barrel, by the simple expedient of driving in a turf-cart past the birds. Golden plover are equally gun-shy, and yet in crossing a bog one may at times flush them from a tuft of rushes almost under one's feet.

In rough shooting, however, skill in woodcraft need not always be exercised. In a little creek where the river is fed by water that trickles down from a bog, over mossy slopes to the still pool fringed with sedges, some wild ducks have reared their brood. There the old birds and the young will have their headquarters until sharp frosts drive them away to warmer estuaries. You may pass and repass their haunt without disturbing them, but if you pause to get a view through the network of branches, or stalk cautiously, intent upon a shot at them, they will be up and away towards a great tarn on the moor. Your best chance is to send a water-spaniel in upon them and wait. Then they will take one narrow circle before shaping a course, as if doubtful about allowing themselves to be driven away by such an intruder, and you may bring one or more down, if your nerve be good and hand ready, as they wheel above the low tree-tops. Now among the larch trees that stretch their slender twigs like a hazy network across the riverside path your dogs may flush a woodcock, who darts in zig-zag flight between the shafts, or with apparently lazy wings goes down the long alley straight from you. There are some men, and good shots too, who cannot hit a woodcock going away from them so. His seeming slowness baffles them, and they would rather take their chance at one darting right or left through the netted branches. If shooting companions are with one—and few of us like quite solitary sport—the cry of "Mark cock" is sure to be frequent in this copse, for there is favourite feeding found yonder in the soft mossy banks of a tiny streamlet beside which holly trees grow.

By some giant boulders in the next glade we may pause for luncheon—a frugal meal—and, after one welcome pipe, trudge on towards the moor. Our path thither leads through a gorse brake, where a friend's fox terriers are of timely use to draw for rabbits, and I know no phase of rough shooting that has more fascination than this. It is well to have a wire-haired terrier for working the runs in gorse brakes or thorny thickets. His size enables him to get through where a spaniel would be frequently caught by the ears, and his pluck makes him regardless of scratches. The one fault that mars his usefulness in other forms of shooting is a merit among the furze bushes, where his freedom of tongue serves as a guide to the way rabbits are running, and if he babbles sometimes, one soon learns the difference between that and a genuine cry. Once out of the brake, that fox terrier must be kept well at heels, or, still better, in leading strings, for his tendency is to range wide with frequent yelps, disturbing every kind of game that may happen to be about.

Now, with spaniels working close ahead, we make our way across a stretch of short heather and reddening branches of the

ROUGH SHOOTING.

whortleberry, where black game delight to feed at morn and
eventide, but it is useless to look for them there now. Partridges
sometimes sun themselves on the next slope. We try for them
in vain, and make up our minds that they will be found later
among the turnips or long stubble on steep uplands along the
coombe, where no reaping machines have ever been seen at work
yet. In the hollow yonder between two waves of brown heather
is tawny moorland grass that looks withered and dry. Let us
work that steadily with guns wide apart like skirmishers, and
always keeping a keen look-out forward. Moorland hares are
wild and fleet of foot. One is certain to start out from a tussock
presently, and if you are not ready to shoot when the chance
comes, there is little hope of bagging her. The next moment she
will have disappeared in one of the furrows she knows so well,
speeding away with ears laid back and legs outstretched, so that
a little ridge serves to hide her from view until she crosses the
crest sixty yards off. But better luck is likely to be in store for
you if the dogs are not too eager. In that grass-grown hollow
which has now the colour of an African veldt in hot summer-time,
I once saw three hares on foot at the same moment, and each fell
before the gun of a different sportsman. A minute later we
started an old dog fox from the lair in which he had been curled
up, sleeping peacefully after his midnight feast, or perhaps
waiting patiently until the rabbits should come out to feed on the
short sweet grass and wild thyme leaves that grow about the
rocky mounds of a neighbouring burrow. There is our favourite
ferreting ground, and we may be sure of a full bag on days set
apart for that sport. Towards evening we may knock over a few
of them, as with white skuts showing clearly in the fading light,

they scamper off to their holes at sound of a footstep. There is a rustle among the grass not twenty yards off, and you see a tiny yellow head with black stripes on it raised above the tufts. Then it disappears suddenly, to pop up again many yards further. It is a stoat hunting some game, and he pursues it with unfaltering instinct. When next the head is shown a clever shot lays it low, and then a rabbit darting across an opening from one remorseless foe falls victim to another not less sure or deadly.

Now we work on up the coombe towards a rushy bog, and cartridges of smaller shot are slipped into the barrels, for we know that presently snipe will be rising in zig-zag flight from beside the little rill that trickles slowly down there. In a few minutes we have bagged all that can be found, and then we turn for the four-mile walk homeward, adding two brace of partridges to the miscellaneous collection as we cross the stubble field, and getting another woodcock among the birches where shadows begin to deepen, so that one can hardly see to shoot. After such a day of pleasant exercise and varied sport the homely dinner of a moorland manor house is better than any feast to which fastidious *gourmets* can sit down in crowded cities.

NOVEMBER.

CHANTREY'S FAMOUS SHOT.

BY OSWALD CRAWFURD.

ON the 20th of November, 1829, at Holkham, in Norfolk, was fired the most memorable shot from a gun that is recorded in the annals of sport. On that eventful day, a little after noon, Sir Francis Chantrey, the sculptor, then plain Mr. Chantrey, the guest of Mr. Coke, of Holkham, was shooting in the woods of that notable sportsman's domain when two Woodcock rose to the great artist. He shot both dead with the first barrel.

The most unsportsmanlike of readers need hardly be told that a Woodcock, rising in covert, gives, by reason of its rapid and twisting flight as it threads the branchage of trees and shrubs, the most difficult of shots. Hardly two Woodcocks fly quite alike. While one will flit silently with the smooth wing-motion of an owl, the next bird that rises will turn and twist with the zig-zag flight of a snipe on the wing. Moreover, the Woodcock is a comparatively rare bird in most parts of this country; few men therefore can reach the perfection of shooting at him which comes of much practice. Then, again, the Woodcock is almost always shot in thick covert, with the chance of the pellets from the gun being hindered or diverted by tree branches and twigs. Often six or seven woodcock or more are put up in a day's shooting late in the year, and what with the real difficulties of hitting them and the nervousness occasioned to the shooters, sometimes not a single one will be brought to bag. To have shot the one cock bagged in the day is an exploit and an honour; to kill two and by the same shot, of course, confers lasting fame. When we add to these difficulties that Chantrey's feat was accomplished with the clumsy fowling-piece of sixty years ago, with a flint lock, with the often unspherical shot of the day and the unimproved gunpowder, the credit of the great sportsman-artist is enhanced.

Sir Francis Chantrey was a frequent guest of Mr. Coke's at that famous centre of sporting Norfolk, Holkham, and a wooded hill close to the house, known till then as " Quarles' New Plantation," was the scene of his exploit. The shooting party consisted of the host, the sculptor, Archdeacon Glover, and the Rev. Spencer Stanhope. " Chantrey," says the latter gentleman, "was standing in the gravel pit just under the Hall. I was standing next to him, but hid from him by the bank formed by the pit. Knowing how keen a sportsman he was, I was amazed at seeing him run up without his gun, waving his hat over his head. 'Two cocks at one shot!' burst from him."

The sculptor not only shot his birds, he carved them in marble, and their monument is of course one of the treasures of the great house of Holkham to this day. The next thing was to procure an inscription, an epitaph for the birds who had attained death and the monumental immortality which the famous sculptor had conferred upon them. Inscriptions were invited from the men of letters of the day, and, as it is difficult to imagine a subject that so lends itself to easy epigram, short poems in various languages—

CHANTREY'S FAMOUS SHOT.

Greek, Latin, Italian, French and English—began to flow in; Wilberforce, Bishop of Oxford, and Lord Jeffrey, Baron Alderson, Dr. Moberley (Head Master of Winchester), Dr. Scott (the Master of Balliol), Peter Cunningham, Lord Tenterden, Dean Milman, Bishop Maltby, Sir John-Williams and Archdeacon Wrangham, are a few of the names of eminent men who responded to the invitation for inscriptions. Lord Brougham is credited with a Greek epigram on the happy shot and the monument to which it led, but all his lordship did was to obtain an epigram from Lord Wellesley. Very few of these distinguished and learned gentlemen had wit enough to get beyond the somewhat obvious point that the man who killed the birds with his gun conferred immortality upon them with his chisel. And it is noteworthy that not one of the many pieces sent in was deemed worthy of being inscribed on the monument. Bishop Maltby's epigram in Greek, to the effect that "By one man's skill both perished, but the life the sportsman took the artist gave," is neat rather than classical, and more classical than correct, for the epigrammatist has to call a modern English gentleman with a gun "an archer;" though to be sure he is outdone by Archdeacon Wrangham, who declares that the sculptor slew them *unâ sagittâ*, with "one arrow"! A better epigram is Dean Milman's :—

"Uno ictu morimur simul uno vivimus ictu."

"By one stroke we died, by another lived"—though, to be sure, it takes a good many strokes to accomplish a marble bas-relief. One of the best English epigrams on these rather obvious lines is Mr. Bacon's, barring the two redundant syllables in the last line :—

"They fly, they fall, by Chantrey's hand they die ;
Yet live, for he to life gives immortality."

The couplet has, for an epigram, a surplusage of ideas as well as of feet, and would perhaps be better thus :—

"The hand that slew, a life-in-death did give ;
Together died we, and together live."

Killing and conferring life is the gist of the epigram of Lord Jeffrey, the great critic and Scotch Law Lord, who finds a fair pun in his brief, and converts it very pleasantly into—

"The sculptor killed them with one shot ;
And when the deed was done
He *carved* them—first upon one toast,
And then upon one stone."

There are two circumstances, however, which none of the epigrammatists took into account. Being in the main bookish men, they probably never knew that to shoot two Woodcocks by design with one shot is as nearly a physical impossibility as anything done with a gun can well be. Men are fabled to have killed two snipe, meaning to, with a single shot, and I can almost believe it, for snipe are often very thick on the ground, and they are found in open country. I have myself known a Spaniard who frequently killed two quails with one cartridge ; but the quail is a bird of the open, too, its flight very direct, like a partridge's, and in August it is extremely abundant in the maize-fields of Spanish Estremadura. The Spaniard would cover his bird as it rose at his feet, and wait till the last moment of its being within range on the chance of another bird crossing the line of flight, and of his hitting both. Sir Francis honestly admitted that the double shot was a "fluke" ; he saw but one bird when he aimed. The other circumstance, which the epigrammatists were too polite to record, or too art-ignorant to perceive, is that Chantrey's carved effigy of the Woodcocks is extraordinarily second-rate as a work of art, so that the poor birds had to lament, not only their sudden death at the hands of a gentleman who was not their proper enemy, an artist that is, not a sportsman, but that the monument which was to perpetuate their fame did them no sort of posthumous honour. All this would probably not have gone unsaid had the epigrammatists not been contemporaries and flatterers. Posterity, if it were now called upon for an epigram, might word it as follows:—

>   Luckless our fate : a doubly luckless lot ;
>   A sportsman carved us whom an artist shot.

TWEED SALMON FISHING.

By George Lindesay.

The angler on Canadian and Norwegian salmon rivers may boast that there is nothing in the way of excitement to be compared to the first rush of a heavy spring salmon on these northern streams, and that the magnificent scenes of nature among which they flow cannot be surpassed; but there are few things more delightful to the true fisherman and true lover of nature than an autumn on the banks of the Tweed, and fortune must indeed be cruel if, during his stay, he fail occasionally to obtain sport of a high kind.

The valley of the Tweed has incidental advantages—apart from propinquity and civilized accommodation—which a man must be cold, indeed, in fancy and dead to all spiritual emotion not to count for something. The rock scenery of foreign salmon rivers may be grander—in point of fact much of it, in Norway at least, is bare and even tame—but where in Canada or Norway can we find a stream so crowded with the deathless memories of great historical events? Not historical associations alone, but every knoll, every hill-side, every ford, and every pool has a record in fiction, or song, or ballad. These things, that the imagination of man has created, engrave themselves on the mind more deeply than even historical events, and the fisherman is a dull man who walks by Dryburgh, and Melrose and Smailholme Tower, unmoved, and past the meeting of Tweed with waters that are famous in song and story—Gala, Teviot, Leader, and Till.

Tweed is *par excellence* a late autumn or early winter river so far as salmon-fishing is concerned, and that is why we have chosen this most famous of Scottish streams—it is English, too, in part of its course, for that matter—for illustration in November.

From one cause or another the spring fishing on Tweed is not of much account; the glorious tints of autumn have appeared among the Border woodlands before the best of the rod-fishing on Tweed is in full swing, and not infrequently the unmistakable signs of early winter have made themselves manifest in what used to be the debatable land between the two sister countries.

Not until the 15th of September are the nets removed, and the fish afforded a clear run to the upper waters, but even then, not having yet acquired the cloud-compelling powers of our American brethren, the autumn rains are often slow of coming, and October, if not November, sees the best sport of the season. With the first "spate" the salmon rush up the river in thousands, and when the waters have subsided in volume, and assumed the clear brown tint so dear to the fisherman's eye, he reaps the reward of his patience in grand sport! The charmingly situated little town of Kelso may be said to be the central spot for the splendid fishing which is obtained on Tweed in the autumn months. Close by is Floors Castle, the seat of the Duke of Roxburgh, and his Grace's water is second to none on Tweed, and immediately below Kelso Bridge is the celebrated pool, "Maxwheel," the subject of our sketch, where the late Duke—than whom no better angler ever threw a fly

—killed to his own rod in one day no less than twenty-eight salmon and grilse.

In addition to the Floorswater, about three miles in length, which includes over twenty fine salmon-casts, the Duke owns the fishing on the right bank of the river for another six miles as far as Carham Burn, which forms the "March" between Scotland and England. Among the many capital casts on this stretch of water there is perhaps the finest salmon pool on all Tweed, "Sprouston Dub." Within twenty miles of the sea, unless the river be very low, the salmon have a clear and easy run through to it, and the angler who is fortunate enough to have the Dub in his day's beat when it is in order, and after the close of the netting season, need go no further—he will have his work cut out. Many is the good day's sport I have enjoyed on this grand salmon-pool in early autumn, when the air was soft and balmy, or in chill October and in bleak November, when the ice particles "crinkled" on the surface of the water, and the frozen drops sparkled on the line and fly, and when the north wind blew.

One day in especial I recall vividly. It was towards the end of October; there had been a series of floods, but each succeeding so closely upon the other that the river never got into real order; it was always more or less "drumly," and although I had been killing two or three fish a day, nothing great "eventuated." At last one afternoon "She" showed signs of clearing, and during the last hour's fishing I killed a couple of salmon and rose several others. There were very evident signs of frost, the glass was rising and the water falling. As we trudged homeward that evening, therefore, Jimmie and I opined that "the Dub" would fish on the morrow.

TWEED SALMON FISHING

Although we were both aware that it was in the highest degree improbable that the fish would take until well on in the forenoon, we were at the waterside soon after 9 a.m. making preparations for what promised to be a great day. A more magnificent morning I never saw; the air was keen and perfectly still, the woodlands stripped of their leaves by the recent storms, and the brown, ferny undergrowth, and all nature beside, sparkled in the rays of a brilliant sun. The way the fish were rolling about the pools was a sight to see, some of them showing the broad silvery flanks of clean run salmon, others again the brown and red of fish that had been in the river some time. The water was perfect in colour and height, and by ten o'clock a gentle breeze sprang up, bringing with it some welcome clouds.

The previous half hour had been spent in trying and selecting a casting-line worthy of the occasion, and, most important of all, in choosing the fly. We examined dozen after dozen of exquisitely dyed and gorgeous insects, *Doctors*, silver and blue, *Durham Rangers*, *Stevensons*, *Dusty Millers* and innumerable other beautiful specimens of the fly-tyer's art, but at last, after a tremendous discussion and a stiff glass of whisky for luck, we came to the conclusion that a double *Jock Scott* was the thing, and armed with my nineteen foot "Forest" and perfect tackle, we got afloat soon after 10 o'clock. In spite of their numbers, however, the fish rose very badly and short at first, and when we got down to our *pièce de résistance*, the celebrated "Sprouston Dub," we had only two fish in the boat. But by that time things had altered for the better, the frost had gone out of the air, the breeze had freshened, and the clouds more frequently overshadowed the sun. At the third cast a good fish rose, and jamming the double steel hard into him, I found myself fast in a lively clean run salmon who made the reel shriek, and did not end his first rush until he had close on a hundred yards of line out. But the hold sufficed, and I held on very hard in order not to waste time, and in ten minutes Jimmie had him in the big landing-net (the use of the gaff on Tweed, from the 15th of September until the 1st of May, has very properly been made illegal), a bright twenty-two pounder. The fun then became fast and furious; fish after fish rose, boldly and well. Some escaped after an intimacy of longer or shorter duration with my fly, but a good many were accounted for, and by two o'clock, when the pangs of hunger began to make themselves felt, we had nine splendid fish and a couple of grilse in the boat. So slow is the current in this fine pool that in order to "bring the fly round" properly it is necessary to begin at the foot. The boat in which the angler stands is then slowly rowed upwards, and the motion thus conveyed assists that of the stream, the line being cast out at a right angle to the boat (which is, of course, kept well out of the fishing-water), and allowed to "come round" until it reaches the near side of the current, where it is permitted to dwell for a few seconds. More than once I lost fish that day by withdrawing the fly for another cast too quickly, only just pricking them; they had followed it in from the far side, and had waited until it assumed a convenient position before they rose at it.

Although the work had begun to tell on my backbone, luncheon was very quickly got over, no interval being allowed for "baccy," and with pipes alight we went at it again. The first fish of the afternoon was a noble fellow of 30 lbs., but not so bright a salmon as most of the others. Then came three others, a grilse and—a catastrophe. The brief autumn day was fast approaching its termination, and we were just shoving off to try our fortune once more, when right out in mid stream I saw the rise of a fish greatly exceeding in size anything we had seen before. Jimmie declared his belief that it was as big as a " coo," and after an extra careful examination of my cast I put up a *Dusty Miller*, in the hope of tempting the huge salmon. But *Dusty Millers* were not apparently in his line; he would not move, and we retired discomfited, only to return to the assault, however, with a *Silver Doctor*.

It was a moment of almost painful interest when the new fly dwelt temptingly over the spot where the huge tail had last been visible, and Jimmie's eyes seemed jumping out of their sockets. Then there was a heavy swirl in the brown water, the line slowly tightened, and I was *into* a real heavy salmon—such a fish as one gets a chance of perhaps once or twice in a lifetime. Down he sank into the depths, and secure in his great strength, took no notice of the vicious " strike " with which I drove the hook into his jaw, or for some minutes of the heavy pressure I immediately brought to bear on him with the powerful Scotch rod. There he lay, some ten or twelve feet beneath the surface, the line humming with the strain, the good greenheart bending double; Jimmie and I, with our hearts in our mouths, awaiting his first move.

It came with a vengeance; slowly at first, and then more quickly, as if gathering way, he began to travel against the stream, until his progress became a mad and irresistible rush, with which Jimmie's oars had no chance. 120 yards of line had been taken off my reel, when at last he came to the top of the water with a mighty roll, and turning, shot off across and down the pool. Then came the disastrous finish. He had been on rather over twenty minutes, when, in the middle of another grand rush, the fly came away, the rod straightened, and all was over! Only one of the double hooks had entered the fish, and the strain had proven too much—it had broken clean off.

Language not being equal to the occasion, both Jimmie and I wisely held our peace; indeed, several days passed before we could discuss the episode with anything approaching equanimity. We pulled ashore, the dog-cart soon turned up, and I returned to Kelso, after one of the best day's salmon-fishing I ever had on Tweed.

HARE HUNTING ON THE BRIGHTON DOWNS.

By H. H. S. Pearse.

For full enjoyment of sport, as followers of the Brighton Harriers delight to describe it, one need be well mounted on a quick and clever horse that can gallop up hill or down, have steady nerves, a firm seat, good hands, and a stout heart. You may perhaps have seen among the horsemen and horsewomen riding out of Brighton towards a fixture on the hills, some who did not seem to be endowed with all these qualifications; but they would be the last to admit that it was possible for anybody to get on creditably with less. Their vivid descriptions of the precipices down which they ride at headlong speed are enough to take a listener's breath away, if he does not happen to have learned by personal observation how small, in proportion to the people who hunt and talk about it afterwards, is the number of those who greatly dare. Pace is, without doubt, a characteristic of hunting on open downs where hares are stout, and there are neither fences, nor furrows, nor tangled undergrowth of bracken, brambles, and sedge, to prevent hounds from viewing their game frequently. Sport under these conditions differs very widely from the slower pursuit across lowlands, especially if among them ploughed fields and dusty fallows prevail so greatly that harriers cannot run a hundred yards in full cry without being brought to their noses. Equally true is it that many followers of the Brighton pack care for nothing but the pace, and ride down steep hillsides at the risk of life and limb as if " the image of war" were worth nothing without its dangers. The majority, however, content themselves with excitement in a much milder form, and keep to the ridges, well pleased if, by cool judgment and skilful skirting, they can cut in occasionally with hounds and enjoy a quick burst; while the bolder few, having dashed madly down the slippery slopes, are toiling slowly up again. Southdown hares are not so timid as to be easily turned from their points or driven into intricate doubles at the sight of horsemen galloping hither and thither; but one can remember how in old days many a good run was spoilt by eager riders pressing too closely upon the scent, or by scores of skirters foiling it in their anxiety to be with the pack for a few brief minutes. One master used rather to encourage wildness by his own example. Mr. Hugh Gorringe has, however, changed all that, and will have his hounds hunted on scientific principles or not at all. Hard riders and skirters alike are kept under due control, and so a day with the Brighton Harriers, instead of degenerating into a series of quick scurries that would make hounds as wild as their followers, becomes a very pleasant experience, in which old-fashioned sticklers for every formality, not less than the lovers of pace, will take delight. If fortunate in the selection of a good fixture, and enthusiastic enough to ride with the hounds wherever they run, a stranger will find his day's sport quite as much as horse or man can get through without fatigue, and neither will care for a parade on the King's Road afterwards, though that is apparently the end to which a good many frequenters of Brighton who don hunting costume direct their ambition.

A fashionable fixture with these harriers is in many respects unlike any other gathering of hunting men and women in England. Not even the Queen's buckhounds on their opening day at Salt Hill attract such a variety of followers as may be seen scampering over the South Down hills in pursuit of health and sport. The rough element is, however, generally conspicuous by its absence, and the riders, if not all distinguished for grace or skill, are on the whole harmless. The Devil's Punch Bowl is one of their favourite trysting places, because there the attractions of hunting and of a monster picnic may be combined. To see the motley cavalcade wending its way over the hills when spring sunshine tempts all Brighton folk that way, one might imagine that the whole population had turned out to keep some time-honoured festival. Along the ridges that command views down precipitous steeps northward and across a wide expanse of green waves in other directions, vehicles of every make and shape are sure to be ranged in line. Round the pack crowds of horsemen gather so closely that the kennel huntsman is kept in a state of perpetual anxiety lest some of his favourites may be kicked or trampled by the restless hoofs, and nobody is so glad as he when the master waves a signal for the hounds to begin drawing. There are some among the hundreds on horseback who do not welcome the call to action quite so joyously. They have begun to realize the possibility of difference between themselves and their impetuous steeds, who, knowing all about this business, are eager to play a conspicuous part in the first rush. Doubts as to whether man or horse may be master are a little discomforting when the chase seems likely to lead down hillsides steep as the roof of a house. The time comes quickly for a solution of such questions. In the nearest gorse brake, or in a patch of turnips beyond, the hare has been lying snugly concealed, yet wide awake to all the sounds of bustle round about. So much on the alert is he that he does not wait to be found by the pack or ignominiously whipped out of his form. Confident in the possession of speed and cunning, he is ready to try conclusions with the enemies whom he has perhaps beaten by some clever ruse more than once before. A March hare is proverbially wild, and the bright sunshine, or a keen, cold breeze in this fickle month, will often befriend him. Once well away, he can show the clamorous pack his heels and bid them defiance for an hour at least, if he does not escape them altogether. But it is not so easy to get away. The first note of hound music, or a shrill "See! Ho!" uttered in quavering treble by some spectator who does not know the rules of the game, is certain to let loose the impatient horsemen in a rush which no mandate of master, or entreaty on the part of his self-elected deputies, can restrain. The timid are carried along in that rush by the bold, and for a time it looks as if everybody were trying his utmost to catch the hare without the aid of hounds. But those who still hold command of their horses find discretion the better part of valour very quickly. The hare, by a dexterous turn, dodges the moving maze of hoofs and dashes down a hillside steeper than many care to risk descending. The pack swings round quickly on the line only

HARE HUNTING ON THE BRIGHTON DOWNS.

just in time to escape destruction at the feet of some hot-tempered steeds whose riders have lost all control, and then there is a sudden halt of the main body, which by one impulse comes to a standstill on the verge of that forbidding declivity. A few bold riders push through the throng, and, with horses well in hand, follow the lead of hounds. The less experienced try going down aslant the slope, until they discover how risky that crab-like motion is; but two or three who have hunted over these downs from boyhood go straight, so that their nimble steeds with cautious bounds may find firm footing at every stride. They go slowly at first, but weight and impetus begin to tell, until every horse seems to roll like a ball, with a speed that carries him far across the rounded fields below. It is not always safe to stay on the ridge in expectation of the hare coming back, as the timid and too clever discover when they hear the chase rolling away from them in the distance. Over a brook and stretches of level meadow, beyond, the hounds speed on, followed by no more than half-a-dozen riders; into a belt of copse, then out again, and run wide rings across enclosed fields, where stiff fences give the fortunate few a merry time. Finding no safe refuge in woodland, hedge, or sheltered hollow, the hare tries at length to make for a haven of rest on the hills, but strength failing before that can be reached, he falls into the jaws of his relentless pursuers, and people waiting on the ridge see only the end of a good run. Not very often does this happen. More frequently hares found in the lowlands take at once to some favourite point on the hills, and if native there they seldom leave high ground except for a few minutes, and with the evident object of shaking off pursuit on the steep ascents as they come back again. There is method in every twist that a cunning hare makes, and if on the South Downs they run straight more frequently than in other countries—sometimes making a point of three or four miles as the crow flies—it is not because they lack the subtlety that distinguishes their race elsewhere, but rather because the Brighton Harriers push their game along too fast for shifts and subterfuges to be of any avail.

DUCK-SHOOTING ON THE BROADS.

By Oswald Crawfurd.

All ducks are grey in the dark, might pass as a proverb with the flight-shooter who practises his sport at nightfall, or with the punt-shooter who works when the deeper shades of night are on him, and at times can see so little of the form and colour of his quarry that he points his great punt-gun towards, and fires at the mere sound of feeding wild-fowl. The shooter by daylight, however, needs to be something of a field naturalist, and to know a mallard or a pintail from a scaup, a smew, or a merganser. If he fail to distinguish his Ducks, he may easily fill his game-bag with some of the most uneatable of winged creatures.

Ducks are divided, not so much by the naturalist as by the sportsman, into two distinct kinds—the diving Ducks, or those which seek their food at the bottom of the water; and the non-divers, or those which only dip their heads beneath the water for food, but not their bodies; or, if they dive, dive but a little way down. The divers seem to find some very questionable food in the muddy depths, for the flesh of nearly all of them is fishy in taste, tough, oily, and unfit for food. Among the divers are the pochard, the golden dye, the different kind of scoters, together with the smew, scaup, and merganser above mentioned. These Ducks are all, as a rule, frequenters of the sea, and all except the pochard nearly uneatable ; while the surface-feeding Ducks—who are, if I mistake not, eaters of vegetable matter only—haunt the sea only occasionally, and love best the river, the reedy pool, and the marsh. The surface-feeding ducks comprise the mallard, or common wild duck, the teal, shoveller and garganey, the pintail and the widgeon—all birds that give excellent sport in the marsh and by the river-side, and make an excellent dish in the kitchen. From this latter point of view, it may be remarked that they probably owe their agreeable flavour, not so much to any intrinsic culinary virtue of their own, as to the fact that they feed on the wholesome weeds that grow in fresh water ; while the non-cookable fowl feed among the fishy, briny slime of the sea-bottom. In proof of which it may be observed that the diving pochard is only good to eat when he resorts to fresh water, and then he is very good ; while the non-diving widgeon loses all fitness for the spit or oven when he goes to sea, as he is far too fond of doing.

There is no easier bird to approach in his inland haunts and none easier to kill flying, and none better to cook and eat than the common Wild Duck. All this is particularly the case if he be found in some such rich inland feeding-grounds as our Norfolk Broads. Here a succession of shallow lakes, rich in all the water-weeds that make good duck-food, are margined with reeds and rushes, and interspersed with patches of land that are seldom dry enough to grow aught but the osier and the alder. Through this paradise for the Duck and his human enemy, the shooter may wander in his punt—passing quietly through the dividing reeds, and entering one after the other of these silent pools. Every here and there he will come unseen into the midst of companies of mallards, teals, pintails and gadwalls ; and get the best of sport and the easiest of shots.

DUCK-SHOOTING ON THE PROAIG.

COURSING.

BY AUBYN TREVOR-BATTYE.

IT is not strange that the votaries of coursing should assert emphatically, as so many of them do, that England holds no other sport so fine. The marvellous speed of the greyhound is in itself sufficient to account for this. It is said—I believe with truth—that the Cheetah (*Chita*), or hunting leopard, can for a short distance surpass the greyhound on a turn of speed. But it is, is it not, matter of history that the famous racehorse, Flying Childers, was beaten by a Greyhound? And even now there are rumours again of a match impending between a champion of either kind. But matches of this sort are very hard to bring fairly off. It is almost impossible, under these circumstances, to ensure your Greyhound running "all it knows." The writer personally will never believe that the best racehorse that ever was foaled could beat a crack Greyhound, if the Greyhound meant to win.

With the exception of the bulldog, whose use has passed away, the Greyhound, as a specialized type of domestic animal, stands quite alone. There is far more difference between a Greyhound and any other form of sporting dog—the deerhound excepted—than between the closest bred racer and the clumsiest cart horse that ever pulled a plough.

Those who are not personally interested in coursing have no idea of the forethought, care, and anxiety that is implied by a dog being fit to win in any company. The breeding itself entails a knowledge of points, structure, performances, and ancestral characteristics, a power of selection partly intuitive and partly learnt, that comparatively few men possess. The treatment of puppies and the training of " saplings " are provinces in themselves upon which as much care is lavished as in the preparation of a prince for the duties of a throne. A pup may have to be brought up " on the bottle," it may have to be reared on Dr. Ridge's food, and very carefully must it be kept from draught and damp, and from anything that would hurt it. It must, as it grows older, have just so much freedom as will develop it to the uttermost, but not so much that it is ever left to look after itself. Above all it must never be chained up, for that will bring bowed legs and many other ills. When the puppies—called saplings till they have passed their first season—are at exercise or training, everything must be avoided that can hurt their feet, such as flinty ground or stubble fields. On the other hand, the more road exercise they can have the better, for this hardens their feet and makes the pads firm. Nor is it wise to let them do much galloping up hill, for even though the gradient be slight this will tend to spoil the dog's shoulders and throw him out of form.

The Greyhound was not always trained with such care as now, for match coursing is a practice of comparatively recent growth. The first coursing society was formed, we are told, by Lord Orford, at Swaffham, in 1776, and this was followed, at varying intervals, by several others, till 1825 saw the birth of the Altcar Club, and 1836 the establishment of the Waterloo Cup.

COURSING.

But the year 1876 was marked by the introduction of quite a new feature in coursing history, and one which it is impossible to leave unnoticed here. I refer, of course, to "enclosed" coursing meetings. It has been said, I know not with what truth, that the idea of the Plumpton enclosure, the earliest made, was suggested by the anticipation of the Hares and Rabbits Bill, and the consequent decrease in the supply of hares that was foreseen. This Bill has been the death-blow to many meetings, but that the enclosed meetings were no wise alternative is seen from the fact that Plumpton, Gosforth, and all the others are numbered with the past, excepting, I believe, two, one of which is Haydock, the scene of Fullerton's early laurels. The objection to coursing in an enclosure rests on two grounds, sentimental and practical. Sentimental, because there is something in the very idea of crowding hares together in a patch of covert and then letting them out one by one to be hunted between wire walls, however wide apart, that is opposed to the breezy spirit of true sport; and practical because, being always over the same change of ground, and always on the whole in one direction, it taught the dogs to run cunning and develop too much calculation. As against this it has, no doubt, tended to give the Greyhound of to-day a finer turn of speed. In the old days—in quite the old days—speed was everything. The dog that first *caught* the hare won. But there was to come a time when reflection showed that if there was one incident more than another in a course that turned on chance, it was the kill, and thus they came to reckoning points. And of these points, as Mr. Thacker says, "a go-bye, a cote, a turn, a wrench, a tripping, a jerking, or a kill of merit may be called the fundamental ones," and so they may be to-day, excepting that "cote" has dropped out, because it was a term of confusion, and "jerking" is lost in the rest.

And now that we are fairly out upon the flats, the story of just one course may show how the points come in, though we must tell it very shortly because of want of space.

Down on an Essex farm a farmer is out with his boys, trying, in view of a local meeting, two young dogs, a brindle and a black. The scene is one of those large grass flats cut up by dykes in all directions, crossed by planks and waggon-ways, and widening down to the tidal marshes where the wild ducks come to feed.

A strong hare is started, and presently the greyhounds are slipped. Somehow the brindle is quicker on his legs, and puts a good clear length between himself and his rival before the other is well under way. But the black is not to be denied, and running gamely, fairly distances the brindle and is soon a length ahead. This gain *in a straight run* is the "go-bye," and scores *two ;* and now the dogs are close upon their hare. And the brindle, running on the right, pushes the hare so close that she swerves a bit away, and the brindle scores a "wrench," in value *half a point*. But that wrench, slight as it is, has favoured the black, who, running finely in, fairly turns the hare at a right angle and almost into the brindle's jaws. This is the "turn," and counts *one*.

And so the chase goes on with varying chances till presently the hare, with the Greyhound on the top of her, leaps at a half-dry dyke. Over goes the brindle and lands clear. The black is not so fortunate, just dropping short to emerge dripping wet and stand with her rival looking stupidly around, for where is the hare? By a device not seldom seen she has jumped short, and doubled in under the culvert of the bridge. And there, if she lies still, the black and the brindle may both go hang, for they are not smell-dogs but only gazehounds, and are now "unsighted," to use the proper term.

There is neither trip nor kill "this journey," for there was no one there to see. Nobody, that is excepting the artist who drew the illustration to this chapter.

ROE SHOOTING.

By J. MORAY BROWN.

THE roe (*Capreolus capria*) is the smallest of the three species of deer indigenous to the British Isles, the male being only somewhat over two feet in height. The colour of the roe-deer varies considerably with the season of the year, being reddish during the summer, and changing to slaty, bluish grey as winter approaches, while the coat becomes very thick.

Shall I describe the haunts of roe-deer as amid the wild, romantic glens of the North, where from the tangled brown heather, vivid green birch-trees, with their delicate, silvery barked stems, uprear their heads; and where rock, mountain, and water combine to render the scenery romantic and vivid? Hardly; for, though you may, and indeed often will, find roe-deer amid such surroundings, you will more often meet with them in sombre pine-woods, or among plantations of young fir and oak, and on the borderland, between cultivation and the barren forest region. There you may often see them feeding on young clover or oats, and if you be a naturalist as well as a sportsman, you will derive no small pleasure from watching their movements, and noting how the fastidious, graceful little deer wander hither and thither, nibbling the tender shoots of the wild rose and bramble that clothe the wood fences; you will often see them play sad havoc with the tops of young oak-trees, a habit that has given them a bad character for destructiveness with foresters. Until the leaves fall, oak woods are favourite resorts of roe-deer; then, once the branches are bared by frost and wind, they quit these for fir plantations, as affording better shelter and more quiet. Roe, like other deer, shed their horns, and it is generally the first week in April before the new growth is clear of velvet. In this respect, however, they vary with seasons, being more backward some years than in others; but in regard to fitness for food, they can hardly be said to be in condition till November, and they are at their best about the end of December and during January.

Compared with deer-stalking, roe shooting is naturally inferior as a sport, for the quarry is more insignificant, and its surroundings are different; yet it may be made a very effective substitute if the habits of the animal be studied, for it will be found in some methods of pursuit a sufficient test of that knowledge of woodcraft which is the very essence of sport.

Roe-deer shooting may be followed under various conditions and in various ways. They may be stalked, beaten for, or hunted by a hound or hounds, and so driven to the guns posted in "passes," for roe-deer, like hares, have their regular runs. In Germany, indeed, another method obtains, viz. "calling" the bucks during the rutting season in July, when the roe forgets and forgoes his generally cautious habits, but this method rightly finds no favour with the British sportsman, and when shot at this season the animal is nearly worthless for food.

Let us now glance at one of the methods of pursuit usually employed in Scotland, for that is nowadays their main *habitat*, though roe are still fairly abundant in parts of Dorsetshire and Somersetshire (I remember seeing no less than seven one day when hunting with the Cattistock hounds a few years ago). As stalking roe-deer will most commend itself as a form of pursuit to him who loves to kill his game unaided, we will describe that form of sport. But it is not to be imagined that there will be any spying the ground, any great number of miles to be traversed, any great exertions, or any stalker to take you up to your game, and, after making you assume every undignified attitude and contortion of which the human body is capable, finally put a rifle into your hand and tell you to "tak' time." Our sport will perhaps be more prosaic, less fatiguing, and yet hardly less satisfactory, and if by our own individual skill and observation we attain our object—to wit, the shooting of a roe-buck with a good head—we shall feel nearly as proud as the slayer of a " muckle hart."

We will suppose you are on suitable ground, a stretch of beech, oak and fir forest, with young plantations trending up to some mountain ridge, where the young trees get smaller in growth the higher the altitude ; the day a bright crisp one towards the end of November, and the afternoon at your disposal. Your weapon a Holland's rook rifle or a Winchester repeater—the latter perhaps for choice. We will suppose you know your ground, and that during the summer and autumn months you will have made yourself acquainted with the habits of your game—noticed their "runs," observed where they go to feed, and that you have marked some trees round which the roe are in the habit of

ROE SHOOTING

playing. Under such trees, the ground will be found cut up and trodden down just as a field is where horses have been exercised. First you will stroll through a stretch of woodland where the dark pines, yellowing larches and russet beeches give colour to the scene; and as you go, you will do well to cast ever and anon a glance upon the soft spots in the ride, in order that you may ascertain if roe are still about in the woods. You can see no signs, so, climbing the wall of loose stones that surrounds the wood, starting a rabbit as you do so, you cross some grass-fields where a few sheep and cattle are feeding, then strike across a furzy, heather-grown common, and finally reach a narrow strip of young fir-trees, which, some five hundred yards further on, joins a large plantation. There will probably be some four hundred acres of covert, and it would appear that looking for roe-deer here would be much like looking for a needle in a bundle of hay. But now your woodcraft comes into play. You will search the runs of the deer, noting if tracks be fresh, and in which direction they lead; then, too, you will perhaps remember a certain patch of wild roses and brambles that you have noted as a favourite feeding spot, and so you will form your plan of campaign. Several passes or runs have been examined with no result, till in a bare, moist spot in the ride along which you are walking a fresh slot arrests your attention. There is no mistaking the sign, for a blade of broad, coarse grass has been partly trodden in, and, from where the roe-deer's hoof has crushed it, the sap is still exuding. Yes, it is fresh enough, and your deer, probably not ten minutes in front of you, is heading towards a patch of oats you know of, green still, and unlikely ever to be harvested. You hurry on, till, as you turn a corner, you come suddenly on a doe and her two well-grown, large-eyed fawns. A moment they stand and stare at you, then bound into the covert, showing the white patch on their quarters as they go, and luckily in the direction opposite to that which you are taking. For half-a-minute you catch a faint sound of their progress, and then all is still. Now you breathe freely; ten to one they will not alarm the buck, but the incident has made you cautious. At length you near the spot where you expect to find your game; the wind, what there is of it, is blowing in your face, and the boundary fence is only some twenty yards in front. Lie down now, and worm your way like a snake through the tangled heather, and underneath the young fir-trees, creep on silently and carefully, and when the fence is reached, peer cautiously through the straggling gorse-bush that surmounts the dyke.

Ah! there is the buck, right out in the open, for most of the crop round the outer edge of the field has been beaten flat by wind and rain. Ever and anon he raises his head suspiciously. But he is a good hundred and fifty yards distant, and stern on. It will not do to risk a shot; he has too good a head for that, and a stern shot is not sportsmanlike. Further on, and to his and your right, the fir-trees grow higher. A deep ditch, too, runs under the dyke, and on the covert side; so you creep back, make a wide *détour*, run up one path and down another, and finally reach the ditch, up which you creep, crouching till you get to a patch of tall heather growing on the bank. Now, look! A glow of satisfaction pervades you as you have ocular proof that your tactics have been successful, your judgment correct; for there, only some fifty yards away, is the roe-buck, still feeding, and with no suspicion of your presence. It seems a sin to slay him, and now that the moment for which you have toiled has come, you would almost wish it had not. But the hunter instinct prevails, the deed has to be done—and then—that fine pair of horns! You push your rifle over the bank, give a low whistle, and as the buck raises his head, aim at his neck just in front of the shoulder and—press the trigger. He makes a wild bound forward and rolls over, dead. Success is yours; those gracefully-shaped, spiked, and rugged horns will be a memento of sylvan sport, and you won't forget the venison, for *côtelettes de chevreuil* with "rowan" jelly is a dish by no means to be despised.

DECEMBER.

DECEMBER SPORT IN THE HIGHLANDS.

By George Lindesay.

The last leaves of departed autumn have long mingled with the dead fern and undergrowth of the woods and coverts; and the rod, which has for so long held its own, has been finally laid aside for the gun. Yet it seems but yesterday when, beneath the shadow of the " Hawks' Rock," I killed the last salmon of the season, with wily old Rob, the cleverest fisherman and poacher in all the country side, to gaff it. The pool was in good order, but owing to the number of dead leaves floating down, not a fish would move; time pressed, I was under a promise to shoot some distance off by noon, and having tried three or four good patterns in vain, was about to give up, when after sundry dives into the recesses of a most ancient volume, Rob fished out an insect of strange aspect. Obedient to the expert's mandate, I proceeded to fish the pool down once more, and at the fifth or sixth throw was fast in a twelve-pounder, which formed no unwelcome addition to our dinner that night, after a hard afternoon's shooting.

But now the salmon are intent on matters domestic; they have ceased to interest, but we have plenty to do without them. The grouse have packed long ago, and are as wild as hawks; but even now, within a few days of the close of the season, odd birds are to be picked up on the moor after a frosty night, when the sun comes out bright and warm; the woodcock, too, have arrived, and are to be found here and there throughout the coverts and coppices, while on the tracts of heather-covered and tussocky bog, snipe are fairly numerous, and duck not unknown.

To-day, Jack and I are going to have a day on our own hook, just to see what we can do in the shape of a mixture; and, accompanied only by Ross, the keeper, and a single gillie, we sally forth as soon as we have swallowed an eight o'clock breakfast. A magnificent morning; yesterday there was a slight fall of snow, and during the night a hard black frost; the snow lies dry—an inch or so deep—upon the frozen ground, and in feathery festoons decks the branches of the pine-trees; beyond the line of dark woods, the blue waters of the Cromartie Firth ripple in the early breeze, and a brilliant sun lights up the wintry scene. Within a couple of hundred yards of the house, and before steady-going, painstaking old Fan, the retriever, has quite got into trim, up jumps a lively coney among our feet, and receives three hasty barrels with apparent enjoyment, but the fourth enables Fan to bring it back to us in triumph, grunting and puffing with delight. Now we approach the line of a tiny streamlet, which here finds its way through the broad strip of open covert we are beating. Fan knows as well as we do that there should be a cock or two about, and acts accordingly. Carefully the old bitch feathers along among the frozen fern and undergrowth which border the little watercourse, her grey muzzle close to the ground, her black and curly tail quivering with excitement—" Mark cock " comes from my left, and one brown bird falls to Jack's gun within twenty paces of where I stand, while another, rising to the report, falls to my first barrel. Two woodcocks for two cartridges, a real good beginning!

Beyond the brook there is a mile or so of broken sandy hillocks interspersed with sundry bare-looking and stunted pines; here the rabbits are numerous, giving us some pretty sport and Fan much excitement, but eight bunnies only are added to the gillie's bag before we reach the strip of thick juniper and tall heather which runs along the foot of the abrupt ascent to the moor. This generally contains a blackcock or two, and as we enter it I have the luck to bring down, with a desperately long shot, a magnificent old bird in full plumage, which Fan retrieves, after a somewhat lengthy pursuit, one wing only being broken. Presently from under her nose there bounds away a great brown hare, looking nearly as big as a donkey and a good deal fatter, but she is not fast enough, and Jack stops her with his second barrel just as she tops a bank of peat. This hill-side, or " face," extends for a distance of about a couple of miles, and is really a capital bit of shooting, but it wants, at least, six guns in line to do it justice, the ground being extremely rough, and requiring close work, so we leave it alone for the time being. Now we have reached the top; before us the brown moors stretch away as far as the eye can reach, and behind us at our feet lies a great tract of " Easter Ross," with its rich farm-lands, its noble woods, and its firth-washed shores.

A pair of very steady setters are now let loose, and Fan takes their place in the leash, not ill-satisfied with her morning's work apparently, and not averse to a rest, for she knows quite well it will not be long before her services are again brought into requisition. "There they go," "Hang the grouse," "Next parish," "Steady,

DECEMBER SPORT IN THE HIGHLANDS.

Don," are the remarks, as a great pack of birds get up at least 200 yards off, and disappear over a heathery brow. As we approach the spot where they had been, the dogs crouch low and come to a dead set close together. " Gone away, Don! Gone away, Paddy! Hold up!" The words are hardly out of Ross's mouth, when up gets a splendid old cock within easy shot, promptly receives two barrels, and is stone-dead before he reaches the heather. Fan is allowed to retrieve the bird, which she deposits at our feet without ruffling a feather, and returns to Ross's heel radiant and quite pleased with her sleek, curly self. In some broken ground we are fortunate in surprising a few single birds, occupied no doubt in the mysteries of a grouse's toilet and enjoying the warmth of the sun ; there we bag four brace by dint of our choke-bores, and curving back towards another part of the "face," sit down to lunch, Jack bowling over a white hare on the way with a regular eye-opener— ninety paces, it turned out.

We feast our eyes upon the beautiful landscape spread out before us, but I fear our chief interests are centred in exploiting the contents of a certain grouse and woodcock pie, and arranging the programme for the afternoon. Half-an-hour is all we can allow ourselves, and at the end of that space of time there is nothing left, either of the pie or of a substantial chunk of roast mutton. A liberal allowance of whisky is handed round, the gillie takes the setters in leash, and preceded by Fan, we resume operations on a clump of very thick covert, wherein pheasants are sometimes found. Very shortly a hen gets up, but we let her off; we get a few snap-shots at rabbits, but the covert is too close, and we only get a single bunny, and are beginning to regret the clemency extended to the hen, when from a thick clump of bracken a great cock pheasant rises with an amazing commotion. Although in the agonies of a struggle through a horribly tough bit of juniper, I let go, and have the pleasure of seeing the greater portion of the bird's tail carried away by the shot, and its owner departing but little the worse; then comes the report of Jack's gun, and I have hardly realized the fact that I have gone through the interesting process of having my "eye wiped," when up gets another "long-tail," which comes down to my first barrel, spread-eagle fashion. Two or three more hens are let off, a woodcock causes the discharge of four cartridges, which do not appear to inconvenience it in the least, and we make our way to the "Bog," an extensive tract of moorland interspersed with patches of turnip, clumps of stunted pine-trees, and bits of marsh.

A snipe is the first thing added to the bag, and having marked down another, I am poking about for it in a very wet bit when a fine mallard rises. My first barrel brings down the duck and puts up the snipe, which falls to my second shot, altogether a useful right and left. The first patch of turnip we enter is drawn a blank, but in the second a covey of partridges get up, out of which Jack knocks over a brace, and a third, that had sat close, later on. Then we drop upon no less than three cock in one spot, but only get two of them, the third, although wounded, succeeding in eluding Fan's best efforts to recover. These are followed by sundry snipe, seven of which, and another splendid mallard, are accounted for in rapid succession. But the short winter's day is drawing to a close, the sun is setting in a frosty haze, and although the gymnastics necessitated by our perambulations in the "Bog" have made us warm, the air strikes keen and chill. We are further warned that it is high time to give up by the outrageous muddle we make over an old blackcock about the size of a house, which rises within very easy shot and gets away scot-free.

And so in the fast waning light we turn toward home, well satisfied with our day's winter shooting, for have we not bagged no fewer than ten different sorts ? And even now all is not absolutely over, for as in deep shadow we are skirting a piece of woodland, a shot from Jack's gun rings out sharp and clear on the still air, Fan rushes forward and we discover a roe-buck with a very decent head indeed lying stone-dead among the bracken. Though not a very big one, our bag shows a mixture that it would be hard to beat elsewhere in Scotland. Four woodcock, one blackcock, nine grouse, two pheasants, three partridges, two ducks, nine snipe, one brown and one white hare, ten rabbits, and last, but not least, a roe-buck. Old Ross for once condescends to admit that we shot decently well, but does not forget to modify his praise by some strong expressions regarding that last blackcock we missed !

A COCK DRIVE IN SCOTLAND.

By George Lindesay.

A MONTH of hard frost had brought in a lot of cock, of whose presence we had been well assured at sundry of our neighbour's shoots, and we determined during the last week in January, with the assistance of these friends, to have a final drive in the Birch Wood.

It is 9 a.m. on the day fixed. All have arrived, some with rather rubicund visages after the early drive through the frosty air, but all keen as mustard. The only dog guest is my friend Frank R——'s inseparable companion, " Abe," a huge, black, curly-coated retriever of great strength, whose unerring sense of smell is only equalled by the extreme delicacy with which he mouths his birds, and his intense good nature. Doses, varying in quantity, of ginger-brandy having been indulged in and pipes lit up, we march to the scene of our day's sport. Everything is as hard as iron. As we tramp along the woodland path our steps ring out with metallic sound. Imprinted on the thin layer of snow, which covers the ground, are the tracks of roe-deer, hares, and of wild birds ; while here and there the hoar-frost, which everywhere sheaths the trees and undergrowth with its glistening white needles, has been shaken off a clump of juniper or faded fern by a blackcock or pheasant. Just as we emerge on to the open piece of rough bog, which separates the bit of covert from the slope on which stands the Birch Wood, the first cock gets up, is saluted by every member of the party with at least one barrel, and departs scathless.

The Birch Wood is a wonderful place for wild game, and especially good for woodcock, but the walking is diabolical. It is a strip of wood pretty nearly a mile in length, planted on a steep slope, the surface of which is covered with huge boulders, rocks, and debris. These are, more or less, overgrown with moss, fern, and juniper, while over all there is a thick growth of birch.

From this it will be understood that progress, even in winter, through this fine piece of covert, is necessarily very slow, and that close and careful beating is required. When a large party is assembled we do it in one beat, but on this occasion "we are seven" only, and must make two bites of it, while even then the outskirts next the moor will not be quite thoroughly worked. Two guns are sent forward to the first ride, accompanied by the under-keeper and a retriever, then Ross, having marshalled us in a line, a gillie between each, gives the word, and we enter the sacred precincts of the Birch Wood. The fun is not long of beginning. The rabbit-holes have been stopped as much as possible; the owners rush frantically about, causing the expenditure of no small amount of powder; several pheasants rise and fall, and within five minutes of the beginning of the beat there comes from my right the welcome cry of "Mark cock," and a single shot from the same quarter brings the bird down. Then from my left there comes the same cry, accompanied by the rapid discharge of two barrels. A brown bird glides swiftly between two birch trees in front of me, and in a twinkling " Fan " has brought me my first cock of the day, to the old lady's infinite satisfaction.

Several shots ahead of us announce that game has gone forward, and that the men in the ride are not idle, while the reiterated cry of " Mark cock " tells that there is no lack of the birds.

On emerging in the ride, preparatory to a fresh start, all show symptoms of a struggle; two of us confess to having come surprising croppers ; the face of one man is bleeding freely, the result of too close contact with a bramble-bush, and everybody is more or less smothered with snow and hoar-frost. These, however, are very minor griefs, and the same two guns having been again sent forward, and the line of guns and beaters re-arranged, we once more start on our necessarily slow and somewhat difficult march.

As I am endeavouring to cross a collection of large moss-covered boulders, the nails of my shooting-boots slip on a bit of smooth rock, and, to the astonishment of my next neighbour, I disappear from mortal gaze in a snow-filled cavity, just as a coney, alarmed at my sudden advent, bolts out of it. With a little help from a gillie, I emerge from my temporary retreat, smothered in snow and with a good many aches about the legs, but thankful there are no bones broken; whisky is applied internally with manifest success, and again we urge on our somewhat erratic career. In this bit the thickness of the covert makes the bird-shooting very hard, and the irregularities of the ground contribute materially to the escape of the rabbits; still we do not let everything off, and several woodcock and a few conies are added to the general bag. The last beat before lunch is expected to be the best, as a good deal of game is believed to have gone on; instead of only two guns, therefore, three are sent forward to take their stand at the extreme end of the wood, and the shrill whistle announcing that they are in position having been sounded, we and the beaters and dogs again advance. This proves a hot bit of shooting, and the walking is better; up gets a great " gollaring " cock pheasant close in front of the beater on my right, and therefore within easy shot of my nearest neighbour or myself. Thinking he is bound to take it, I refrain from shooting, so does he, and away goes the bird back over our heads in triumph. " Why on earth didn't you shoot that pheasant ? " bawls one. But two more of the escaped one's friends getting up at that moment pretty handy, render reply to this question unnecessary for the time, and the argument is deferred, *sine die*. As we get on to the end of the wood, a number of cock are flying about, a good many rabbits are crossing and re-crossing the wood, and the firing from the men with the beaters is very sharp ; but I notice that both fur and feather show a strong disinclination for the open, and for the most part break back or upwards to the unbeaten portion of the wood. The consequence is that when at length we emerge from its depths we are saluted with sundry inquiries and remarks by the three sportsmen who had been placed as stops. " What the deuce had we all been firing at ? " " A nice noise you've been making, certainly ! " " I suppose you'll want more carts ! " and so on.

But the morning's work has just put a keen edge on our appetites which will admit of no delay, and we immediately proceed to a small " bothy," where on such occasions as the present we are in the habit of indulging in the luxury of a hot lunch. There is a gigantic fire of turf burning in the little cottage, on which rests an immense iron vessel, crammed with Irish stew, the finest dish possible for luncheon on a cold winter's day, and of that Irish stew, miraculous though it may seem, not one little bite or sup will remain in half-an-hour! No one thinks of drinking anything but whisky; indeed, it would be a farce to suggest any other fluid, and of this there is a supply which, anywhere but in Scotland, might be considered superabundant. I much doubt, however, if any goes back home again. Then, just when we are feeling a bit lazy, and the pipes are in full blast, it is announced that time is up. Like their masters, the dogs stretch their stiffened limbs, the keepers and gillies have a final nip, and soon all are in their places for beating the upper portion of the Birch Wood, back to the end where we began in the morning. There is a general consensus of opinion that, instead of sending guns forward, the line should be extended by one making his way along the rough face above the wood and another along the extreme edge thereof. For this last duty I am detailed, and an enthusiastic sportsman belonging to the Clan Mackenzie scrambles up the rocks and takes up his position on my right.

But once more the fun becomes general. Woodcock dart swiftly among the trees, while rabbits fly for refuge among the rocks on the face, and are freely slated by my neighbour and myself, while an occasional rocketing pheasant gives variety to the bag.

By the time the final beat is ended it is close on four o'clock, and the light is fading fast; but there is enough to admit of an inspection of the game killed. The number of head will not bear comparison with the cartridges discharged; nevertheless, the bag is not a bad one, composed, as it is, of forty-three woodcock, thirty-three pheasants, two hares and eighty-four rabbits. All agree that it has been a capital day's sport.

'LONG-SHORE SHOOTING.

BY OSWALD CRAWFURD.

SHORE shooting is a form of sport with the gun, not so fashionable as the grouse or partridge drive, or as the battue "shoot;" not so sociable, and perhaps not so exciting as rough shooting in marsh or woodland; not so costly, not so murderous, and not so productive of rheumatism as punting after wild fowl, but a sport that requires qualities of eye and hand, and endurance, and a keen sporting instinct as great as these sports require, and a knowledge of natural history far greater. No account of Shore Shooting is to be found, so far as I am aware, in sporting books—none, I believe, in the excellent and fashionable series edited by the Duke of Beaufort. It is not the pursuit of the idle, or the rich. Shore shooting may be pursued all through the autumn months, but in mid-winter it is at its best. Many of the smaller shore-birds, such as the dotterel, and the ringed plover, the redshank, and the sandpiper—are chiefly summer or autumn visitors, and afford fair, if trivial, shooting along our sea-shores during the period of their visits; but with the strong frosts of late October and early November, when the sea is curling with the first keen winds of winter, and changing its summer hues for the slatey grey reflected from the lowering skies, then the 'long-shore shooter knows that in his solitary walks between sea and land he will meet with bigger and better fowl.

No inland region has half the variety of bird-life to offer to the sportsman that he finds on the strip of sand and *salting*, often not two hundred yards in width, which lies between low-water mark and the first tussocks of turf that grow where sea-sand or shingle stones meet the alluvial inland soil.

The shooting itself is partly in the nature of stalking, partly in that of flight shooting; partly the birds rise to the gun like snipe in a marsh, or partridges in a clover-field. There are constant surprises, too, in shore shooting, and the bird that the shooter leasts expects to see often rises at his feet. When he has used his field-glass, perhaps in vain, to find a flock of curlew or whimbrel on the miles of smooth sand, a belated knot or godwit will rise from the rocks between the shooter and the sea and give him an easy cross shot at forty yards; or a brace of teal will start at his feet from a rushy, fresh-water drain as he crosses it, and afford the prettiest of double shots. Then again the path through the air for all the shore-birds is along the narrow strip between land and sea afore mentioned, and the 'long-shore shooter comes in for the benefit of such fowl as are ill-advised enough to fly within forty or fifty yards of his head.

The chief expectation of the shore gunner is to stalk and kill such of the larger waders as he can descry with his field-glass on the wet sands between high and low water-marks. Of these the most sought for is perhaps the curlew—the great, grey, long-legged woodcock of the barren moorland and barren sea sands. He is a bird much in the mind of the shore shooter, less often in his game bag, for he is as wary as a much stalked "royal" stag in a Highland deer forest, and he must be approached by such arts as the deer-stalker uses. The shooter must utilize every inequality of ground or jutting rock, and be very particular to come near his bird up, not down, wind.

When the inland resorts of many of our winter migrants are hardening with frosts, and the sea is tossed into "white horses" by a strong north-easter, the meeting-place of land and sea, being least affected by cold, is the resort of many birds; waders, divers, wild duck and wild geese, that in more temperate weather keep to the hillside or the valleys of the interior. Then again, the purely deep-sea birds such as the brent geese, that in most districts are more numerous than all other kinds of wild goose put together, that hate dry land, and go far to sea to seek their rest, even they are at times driven in by stress of weather, and seek refuge in sheltered inlets and on the oozy flats. This late autumn and early winter season, too, is the migratory time for the larger wild fowl from the north, many of which make their abiding-place in the estuaries and in the great marshes that neighbour the sea. Then again, every change of wind or weather causes the shore-birds, and especially the many kinds of duck or geese, to change their quarters from land to sea, and *vice versâ*, and at these seasons the shore shooter enjoys many a chance of an unexpected shot—not counting that twenty minutes before dark comes on, when the water-fowl pass regularly over certain parts of the coast from the offing to their feeding-grounds on shore.

Most shore shooters work alone. The true shooter, like the true angler, or the true philosopher, loves solitude, and if he tolerates a companion, it is not for his society but for his co-operation. A companion whose room is not twenty times better than his company must be silent and discreet. He should carry unselfishness and obedience, even to the point of subservience. He should be prepared at a moment's notice to make a circuit of half-a-mile for the chance of flushing a curlew and sending it back over his friend's head. I am not sure that he should not carry compliance to the point of going up to his waist in the surf and risking a cold, or a watery grave, to retrieve a wounded duck. If a man won't perform this trifling service for his friend, a dog will, and on the whole, therefore, dogs are to be preferred as companions in 'long-shore shooting. A really well-trained retriever that will boldly face a wintry sea, that will never leave his master's heels till he is bid, and that will crouch on the sand and stay where he is told while his master makes a long *détour* in pursuit of game, is worth all the biped companions in the world.

As for 'long-shore shooting, some people use long, single-barrelled guns, muzzle-loaders, of No. 8 calibre, with a heavy charge. When a man does nothing but stalk sitting birds, this slow and old-fashioned weapon may serve his purpose. With such a gun I have seen some surprisingly long shots made; birds killed dead perhaps sixty or seventy yards off, or even more; but if a man wants to take all comers in the shape of fowl as they fly from, over, or across him, he must be more abreast of the time and use a strong No. 12 double breech-loader, weighing not less than 7¼ lbs. The gun should have a moderate choke say of twenty thousandths of an inch, and as the effect of such a choke is to crush the shot against the sides of the barrel and destroy its spherical shape, he must use hard shot. The shot should be

No. 5 and the charge a full one, say $3\frac{1}{2}$ drs. of black powder, or 50 grs. of Schultze, or of E. C. with $1\frac{1}{4}$ oz. of shot. This is a heavy charge, and it takes a heavy gun and a good one to stand it without unpleasant recoil. Let it be remembered that to obtain a good close "pattern" at fifty yards the case of the cartridge should be only slightly turned over—just enough to keep the wad and the shot in position. Another hint for shooting in very cold weather may be useful. When a man is warmly and thickly clad for winter, he will find that the gun he shot well with when he wore a thin coat in August or September will no longer come easily to his shoulder. He is awkward at snap-shots, and misses his birds. Let him have a quarter or half an inch taken off the heel of his gun-stock and he will shoot as well as ever. This, of course, only if he can spare a gun for cold weather shooting alone.

GAMEKEEPERS.

By Aubyn Trevor-Battye.

"'Well, my Lord,' says Cox, 'I'll do my best; but the foxes, you see, my Lord, kill a deal of game.' 'But you are not to kill the foxes, nevertheless!' says my Lord. 'By no manner of means, my Lord; on'y you see they ain't *always* at home; foxes will travel, and——' 'Cox,' says his Lordship, quite solemn-like, 'listen to me: No Fox, no Cox! Good-night, Cox.' And the gorse was never without a fox after that."

You remember where that comes from? Of course you do. From one of the most charming books that ever were written. The laconism expresses very fairly this general truth, that in an average sporting district you have a right to expect an ample supply of game and foxes, for the keeper who knows his business can give you both. If this is so—and it is written quite deliberately—the man who does not feel that he is master of the subject may be wise to take refuge in this general principle: *Ask no questions, but judge by results.*

It is probably true that there is no class of English servants that has its employers under its thumb to anything like the same extent as the gamekeepers. The reason for this is not far to seek. For of all those who employ a gamekeeper, how many have themselves any practical knowledge of the things that belong to his craft? How many could undertake the keeper's duties for a single week? Only, we suspect, a very small proportion. The butler, the coachman, the gardener even—in a majority of cases —can do very good work by rule of thumb. But the gamekeeper, to do good work, must be a business-like and methodical student of nature, with trained habits of observation and a power of rapid deduction. He must be, then, a man of intelligence above the average. But, as it is, only too many of them just shuffle through the seasons, for only too many are recruits from the ranks of the shiftless loafers. Let us take the seasons through and begin with the rearing of birds.

There are all the duties of the pheasantry. The seeing that all conditions are maintained that are most favourable to laying, e.g. proper proportion of sexes; the freshness of the ground; the right number of birds in each compartment, where a bird or two too many may spoil the whole; just so much food and of such a kind as will keep the birds in perfect health, but will not check their laying, as so easily may happen if they grow too fat; the deciding which nests of the wild birds are in dangerous places, and had better be removed; the keeping a good healthy stock of fowls of the right size and sort for foster mothers; the careful moving of the sitting hens off their nests for feeding purposes every day, and the seeing that the eggs are not over-dry. And then, when the birds are hatched, anxieties multiply; for however clean he may keep the coops, and how often soever he may shift their position, there are still the chances of pip and gapes to be met and fought, and this is generally a losing game. The gamekeeper must try to believe—though few of them will—that as the gape worm, after the wont of Nematode worms, makes a "host" of some worm or mollusc, it is important that the coops should not be in damp situations, and, more important still, that all bodies of dead birds should be *burnt*. Then there is ants' "eggs" collecting to be done; the gentle to be reared; the fox and sparrowhawk to be defeated, and many a small duty that the day's work brings which we have no time to notice here.

When the birds are run off into the coverts they still need a careful eye. Apart from the regular feeding, corn stacks should be placed at points in the woods, and the keeper, as he goes his round in the morning, will be able to tell within a little whether he is losing birds or no. He must not leave these things to anyone else; he must see to them himself. He may find that the wood pigeons are stealing most of his maize, that a stoat is running the bank, or a strayed cat living in the covert. They must be stopped, and this brings us to the question of trapping.

The keeper who is not an accomplished trapper is of little use, for trapping forms one-half of his business. Cats, rats, rabbits, stoats, all have to be taken by a different method; the cat in a trap bushed and baited, the rat in an unbaited trap concealed in a run, the rabbit in a wire, the stoat in a drain pipe, and so on. The trap is the gamekeeper's weapon, and not the gun. Indeed, it is quite an open question whether he should ever carry a gun at all.

That keeper who was not blinded by prejudice, and who was not a slave to the common habit of generalizing from a single instance, would come very soon—I venture to believe—to class together his enemies, the vermin, under two heads in his own mind:—

A WOULD-BE POACHER.

First.—Inveterate foes.

Second.—Occasional foes, or possible friends.

In the first group he would probably place the cat, fox,—a cunning and inveterate poacher—crow, sparrowhawk, magpie, stoat, rat.

In the second, the tawny owl, barn owl, rook, jay, kestrel, weasel, hedgehog, badger.

The writer, at any rate, has very good reasons for so dividing the vermin; but, for want of space, he cannot give them here. But to take just one instance. Of all the enemies the keeper has, not one is more persistent, more insidious, or more deadly than the barn rat. And yet many keepers, who will expend a vast amount of trouble in catching owls and jays, either under-estimate the rat's power or are content to accept him as an unpleasant but endurable fact. He is in truth the deadliest enemy of game.

There is one little trick the keepers are much inclined to try on. It is to leave crows, sparrowhawks, and such like until they have young, and then to kill the lot, and so to make a good show in the keeper's larder. By that time, needless to say, all the harm is done. Few keepers can keep a good dog—or, rather, keep a [good dog good—and fewer still can make one. It is a question of tact and temper. And of all the things that go to spoil a day's shooting these two are the worst: the dog that runs in and the keeper that rates or punishes him.

The keeper must have in him something of generalship, for it is he who has to order the management of the "shoot." He must be possessed of discretion and of pleasant manners, or your shooting may be ruined by the tenant farmers. The keeper must be a man of strong moral character, for his temptations are great, and are often made greater through the fault of his master.

So far we have but touched the fringe of a very interesting subject. But now that we have seen a little of what "keepering" means, it is not strange that the perfect keeper should be a very rare fact indeed. But the well-intentioned, the hard-working, and the obliging keeper is not rare. We each of us know many such in our own little sporting circle. Without the least wish to be didactic, we may be allowed, perhaps, to recommend the following suggestions as the results of long experience :—

It is seldom wise to employ a keeper in his native district.

It is seldom wise to bring a man from north to south, or *vice versâ*.

It is never wise to keep a man, however "sober," who frequents the beershop.

It is never wise to let the local dealer buy your game or rabbits.

And lastly, when one thinks of the keeper's immense opportunities for studying natural history, one is strongly tempted to add this precept: insist upon your man keeping a general note-book.

PIKE FISHING,

By George Lindesay.

It has constantly been maintained, by learned and unlearned alike, that the pike is not a native of these islands, but imported probably by the monks in the reign of Henry VIII. The date of the introduction of this great and. greedy fish is even fixed at 1537, and one common form of a popular couplet is that

> "Turkeys, carp, hops, pickerel and beer
> Came into England all in a year."

If the pickerel, or pike—the fish has as many *aliases* as a burglar —is indeed a naturalized alien, he has certainly made himself thoroughly at home, for there is no one of his compatriot fishes, save perhaps the perch, with his prickly dorsal fin, that the pike does not dearly love and eagerly feed upon.

The learned have now "changed all this." The pike, they maintain, is as much a native-born British fish as the eel or the gudgeon. The *savant* has discovered his pre-historic bones in abundance in the marshes near Ely; though this indeed may only prove that he lived once, may have grown extinct with the cave-bear and mastodon, and has been reintroduced by the monks in the Middle Ages. The late introduction of the pike into England has been so often asserted by the manual-making angler, that it is odd that these industrious gentlemen have never taken a walk to the British Museum and consulted the most ancient of all manuals on the fisherman's art, "The Boke of St. Albans," by Dame Juliana Berners. This famous work was published by Wynkyn de Worde in 1496, and the pike was evidently then looked upon as a native. "The Pyke," she says, "is a good fysshe, but, for he devoureth so many as well of his own kynde as of others, I love him the lesse. For to take hym ye shall doe thus," &c. A hundred years before this the pike is mentioned by Chaucer as a common and highly-valued fish; but as he calls him by the name Luce, which, so far as I know, is not now used in any part of these islands, and was, perhaps, nearly obsolete in Shakespeare's day, the passage has doubtless escaped the eye of the manual writer. In speaking of the Frankelyn in the "Canterbury Tales," the poet tells us that

> "Full many a fat partrich had he in mewe,
> And many a brem and many a luce in stewe."

The pike is the salmon of sportsmen who want means or leisure to hire fishings in Wales, Ireland, and Scotland, or rivers in Norway. He is not nearly so good to eat as any one of the Salmonidæ, though Isaac Walton maintains that a pike baked with a pudding in his belly is a dish for a king; nor does he show such sport as the salmon when he is hooked. Indeed, he often hangs on the line like a water-logged piece of tree trunk, but this is probably because we set about to catch him with a complication of great hooks enough to paralyze the energies of a shark. On the other hand, he has none of the caprice of the salmon; where he lives he abounds, and an hour's journey by train from London will land the fisherman in good pike waters.

PIKE FISHING.

Of the numerous streams where pike are to be found, there are none where he makes himself more thoroughly at home than the Thames, and it is in the waters of the metropolitan river that we Londoners mostly seek his acquaintance. A Thames jack is no fool, and, especially when he has arrived at years of discretion, he wants a lot of taking. In the days of my youth I can recollect fishing for pike with gimp about as thick as my little finger and other tackle to correspond. It was considered quite fine enough, and the fish apparently accepted our theories on the subject implicitly, allowing themselves to be captured with engaging ease.

Not so the Thames jack of to-day; he must be lured with all sorts of delicacies of the most elaborate kind. These must be mounted with scientifically arranged hooks, and the trace must be of single salmon gut. It would almost seem, indeed, as if the time were not far distant when the proper and only thing to use will be trout gut, so great is the objection that Thames pike display for anything but fine gear. *Apropos* of gut, I recollect some years ago killing, on a well-known Scotch loch, when trolling for trout, a pike which weighed over twenty-five pounds on a trout gut trace and with a fourteen-foot rod.

The season for pike begins on the 16th June, but this date is much too early to begin pike-fishing on the Thames. The fish recover condition after spawning very slowly, and it is the opinion of many anglers that September is quite soon enough to try for them. From October to the end of January good sport may be obtained, according to weather. Moreover, during these months the pleasure traffic ceases, and fish and fishermen are not being constantly disturbed by the ubiquitous steam launch. If there has been a flood, too, the weeds have rotted away, and the fish are to be found in the deeper waters in search of food. On a rough boisterous day in November or December the angler, whether from boat or punt, is pretty sure of sport among the jack on any well-chosen reach from, say, The Bells of Ouseley, at old Windsor upwards ; and later on in the year I have had some of my best sport when the frost has been so hard that the ice extended several feet from the banks, and when, clad in the warmest garments and hard at work, it was no easy matter to keep up the circulation. On such a day I have seen jack take greedily, more especially if the water chanced to be a bit discoloured, and the sun coming out induced some warmth during the middle hours of the day.

Spinning and live baiting in their various forms are, of course, the recognized methods of taking jack in the Thames. I am not sure, however, whether these might not with advantage now and then be supplemented with the fly. In open water in Scotland I one day landed thirty-four small jack of from 4lb. to 7lb. apiece with the fly. If the fish can be induced to take the fly it has many advantages ; no baiting (possibly with fingers half-frozen) is required. The casting is as for salmon, and with a salmon rod. The fish undoubtedly give better sport than when they have got a whole flight of hooks down their throat, and they don't mind coming back half a dozen times after being hooked.

The rise, too, is to me a very attractive part of the performance, and no bad substitute for that of a salmon. When the fisherman seeks to lure the jack, winter has thrown her grey mantle over the scene ; the biting wind sighs in melancholy fashion over the surface of the silent river, and rustles among the reeds and the leafless woodlands. Nevertheless, although the luxuriance of summer and the brilliant tints of autumn are wanting to the landscape and the song of birds is no longer heard, Nature is never without some compensating charm to her lover. For him there are inner beauties in the reaches of the Thames beyond those that lie in scenery, even when the land is frost-bound and the January north wind blows. The good angler is always more or less independent of weather or scenery ; and the chances of taking a heavy fish are quite enough for him, even were the scene as disconsolate as the shores of Acheron, or the banks of the Manchester Canal.

THE END.

LONDON:
GILBERT AND RIVINGTON, LD.,
ST. JOHN'S HOUSE, CLERKENWELL, E.C.

www.ingramcontent.com/pod-product-compliance
Lightning Source LLC
Chambersburg PA
CBHW030004240426
43672CB00007B/820